to save the land and people

to save the land and people

A History of Opposition to Surface Coal Mining in Appalachia

Chad Montrie

The University of North Carolina Press

Chapel Hill & London

Designed by April Leidig-Higgins
Set in Minion by Copperline Book Services, Inc.
Manufactured in the United States of America

The paper in this book meets the guidelines for permanence
and durability of the Committee on Production Guidelines
for Book Longevity of the Council on Library Resources.

Library of Congress Cataloging-in-Publication Data
Montrie, Chad. To save the land and people: a history of
opposition to surface coal mining in Appalachia / Chad
Montrie.
p. cm. Includes bibliographical references and index.
ISBN 0-8078-2765-7 (cloth: alk. paper)
ISBN 0-8078-5435-2 (pbk.: alk. paper)
1. Strip mining—Environmental aspects—Appalachian
Region. 2. Strip mining—Appalachian Region—Public
opinion. 3. Public opinion—Appalachian Region. I. Title.
TD195.S75M56 2003   338.2'724'0974—dc21   2002010880

cloth   07 06 05 04 03   5 4 3 2 1
paper   07 06 05 04 03   5 4 3 2 1

To my mother

contents

illustrations and maps

Illustrations

Maps

acknowledgments

Like other authors, I received the assistance and encouragement of many people to complete this book. Most recently, Sian Hunter, Adrienne Allison, Pamela Upton, Laura Cotterman, and the other folks at UNC Press steered me through the publishing process with equal measures of good humor and professionalism. Earlier on in the research and writing, I was assisted by numerous helpful archivists and library staff, including Lyle Brown at Marshall University, Denise Conklin at Pennsylvania State University, Eileen Mountjoy at Indiana University of Pennsylvania, Bill Marshall at the University of Kentucky, and Shannon Wilson at Berea College. During the hot but productive summer of 1998, Richard Cartwright Austin, the late Joe Begley, Ken Hechler, Frank Kilgore, and Sue Ella Kobak gave me hours of their time for interviews, and Louise Dunlap talked to me at length a bit later by phone. Throughout the various stages of putting together a dissertation and then transforming it, Warren Van Tine, Raymond Dominick, Leila Rupp, Loren Babcock, John Cumbler, Steve Fisher, and Susan Freeman read and commented on numerous chapter drafts.

As for my family, my daughter Phoebe has patiently awaited the completion of this book and I hope she finds some small compensation in its contribution to the larger project of creating a better world. Although my sisters did not worry themselves much about what I was writing, they were a source of support for me when things got hard. Then, too, without the many sacrifices of my mother and stepfather, graduate school and the book would not have been possible.

abbreviations

ACSL	Allegheny County Sportsmen's League (Pennsylvania)
AFL-CIO	American Federation of Labor and Congress of Industrial Organizations
AGSLP	Appalachian Group to Save the Land and People (Kentucky)
AMC	American Mining Congress
APSO	Appalachian People's Service Organization
ARC	Appalachian Regional Commission
ARDA	Appalachian Regional Development Act
ARDF	Appalachian Research Defense Fund
ASIS	Appalachian Strip Mining Information Service
AVS	Appalachian Volunteers
CAD	Congress of Appalachian Development
CASM	Citizens to Abolish Strip Mining (West Virginia)
CCASM	Concerned Citizens Against Strip Mining (Ohio)
CCOP	Concerned Citizens of Piney (Tennessee)
CCR	Community Council for Reclamation (Ohio)
CESJ	Citizens for Economic and Social Justice
CLPSR	Citizens' League to Protect Surface Rights (Kentucky)
CODE	Citizens Organized to Defend the Environment (Ohio)
COPE	Cloverfolk Organization to Protect the Environment (Kentucky)
CORA	Commission on Religion in Appalachia
CPOPMA	Central Pennsylvania Open Pit Mining Association

CSG   Council of State Governments

CSM   Council of Southern Mountains

CTFSM   Citizens Task Force on Surface Mining (West Virginia)

DNR   Department of Natural Resources

DOI   Department of Interior

DSM   Division of Surface Mining (Tennessee)

DSMR   Division of Strip Mining and Reclamation (Kentucky)

EPA   Environmental Protection Agency

EPC/EPI   Environmental Policy Center/Environmental Policy Institute

FOE   Friends of the Earth

ICOA   Independent Coal Operators Association (Kentucky)

IWL   Izaak Walton League

KCC   Kentucky Conservation Council

KFTC   Kentucky Fair Tax Coalition/Kentuckians for the Commonwealth

KUAC   Kentucky Un-American Activities Committee

MFD   Miners for Democracy

MSWV   Mountaineers to Save West Virginia

NAS   National Audubon Society

NCA   National Coal Association

NCASM   (national) Coalition Against Strip Mining

NCPC   National Coal Policy Conference

NREP   Natural Resources and Environmental Protection Cabinet (Kentucky)

OEO   Office of Economic Opportunity

OSM   Office of Surface Mining, Reclamation and Enforcement

PCCA   Pike County Citizens Association (Kentucky)

PFSC   Pennsylvania Federation of Sportsmen's Clubs

SCEF   Southern Conference Education Fund

SMAWV   Surface Miners Auxiliar of West Virginia

SMCRA   Surface Mining Control and Reclamation Act

SMSC   Strip Mining Study Commission (Ohio)

SOCM   Save Our Cumberland Mountains (Tennessee)

SOK   Save Our Kentucky

SOM   Save Our Mountains (West Virginia)

SRI   Stanford Research Institute

TCWP   Tennessee Citizens for Wilderness Planning

TVA   Tennessee Valley Authority

UMW   United Mine Workers of America

VCBR   Virginia Citizens for Better Reclamation

VISTA   Volunteers in Service to America

WVSM(R)A West Virginia Surface Mining (and Reclamation) Association

to save the land and people

Common People and
Private Property

One August night in 1968, four men drove onto a strip mine site owned by the Round Mountain Coal Company in Leslie County, Kentucky. They shined a flashlight in the eyes of the lone watchman, tied him up, and drove around in his jeep for four hours, quietly and expertly setting the company's own explosive charges. Just before sunrise, they removed the guard to a safe place, detonated the charges, and left behind the smoking hulks of a giant diesel shovel, D-9 bulldozer, auger, conveyor belt, three hi-lifts, a truck, three generators, and one jeep. Altogether, property damage totaled $750,000. Detective J. E. Cromer, of the state police force, described the destruction as the most extensive he had ever seen in eleven years of investigating sabotage. Yet company vice president Bill Arnold was supposedly dumbfounded about why anyone would go to so much trouble to halt the mine's operations.[1]

The men responsible for blowing up equipment at the Round Mountain strip site were never found and we cannot conclusively determine who they were or what their reasons were for engaging in sabotage. But we can speculate about their identities and motives by setting the demolition in its historical context, which is one of social ferment. During the 1960s, eastern Kentucky and other parts of Appalachia saw a surge of grassroots militancy and movement organizing, including a movement to abolish surface coal mining. The Round Mountain saboteurs were probably part of that collective effort, perhaps even members of the Appalachian Group to Save the Land

and People. In all likelihood, they were attempting to stop operations at the Leslie County mine to send a message to state and federal lawmakers that destruction of company property was the alternative to a legislative ban.

Tensions had to be high for the men to take such drastic action. But how did tensions get to that point? What was so objectionable about surface coal mining? Why would people use industrial sabotage to fight it? These questions have only incomplete answers. Historians have given scant attention to strip mining, and they have shown even less concern with the twentieth-century campaign to outlaw it. The 630-page interpretation of the post–World War II environmental movement by Samuel Hays gives surface coal mining and its critics all of three pages. John Opie's environmental history textbook, *Nature's Nation*, draws on some of the rich sources for Appalachian studies and communicates a better understanding of both stripping and its opponents. Yet his is also a spare overview. Other histories, such as Duane Smith's two-century survey of coal mining and Richard Vietor's analysis of the coal industry's role in politics, also come up short. None of the narrower studies provide either a satisfactory history of the stripping industry or a comprehensive history of abolition efforts. Likewise, the scholarly literature addressing the use of natural resources in Appalachia as well as the work examining the many social movements of the region both fail to explain the evolution of surface mining and efforts to outlaw the practice.[2]

This lack of sustained interest in stripping by historians is certainly not a reflection of its social or historical significance. Surface coal mining has dramatically impacted communities in the Appalachian coalfields. As the industry expanded in the years after World War II, it exacerbated the poverty and chronic unemployment of the region, compounding the impoverishment that was a legacy of other extractive economic activities, including deep mining. Despite their many protests that regulating or banning surface mining would take away needed jobs, strip operators themselves were one of the primary threats to various livelihoods in Appalachia. Their destruction of arable farmland did much to undermine farming as an occupation, while the relative efficiency of strip mining as compared to underground mining allowed for the employment of far fewer miners per every ton of coal mined. Yet it was not only the ruin of good cropland and the small payroll at the mines that pinched off economic development and growth in the postwar years. Under-assessed coal reserves and paltry property tax payments during and after strip mining further hampered local people's efforts to build and maintain schools and roads, and to provide various public services. Operators also balked at returning some of their profits to the region through severance taxes, and local and state infrastructures suffered accordingly.

Surface coal mining affected the environment of the Appalachian coalfields too. Stripping denuded millions of acres of steep slopes and rolling hills in the coalfields, and this loss of vegetation caused soil erosion as well as increased surface runoff. Erosion led to the siltation of streams, and this devastated aquatic life. Increased surface runoff caused heavier flooding and floods where there had been none before. The bare hills also deprived numerous animal and plant species of habitat. Acid mine drainage, produced when sulfur-containing compounds such as pyrite and marcasite are exposed to air and water, polluted streams and groundwater. Even when limestone was present to neutralize some of this acidity, the drainage and acid-laden soil made revegetation and post-mining crop production nearly impossible. In areas where surface mines perched above homes, schools, and whole towns, the lack of revegetation and the abandon with which operators dumped "spoil" down steep slopes led to disastrous landslides. Some of these slides were fatal, burying people alive, while others simply swept away any structures in their path. In addition, blasting at surface mines cracked the foundations of people's homes, sunk their wells (which in some cases were already muddied or fouled by acid runoff), and sent "flyrock" hurtling dangerously into the air.

The poverty and ruined hills caused by stripping are reason enough to make it a worthy subject of close investigation. A practice so devastating and yet so common needs to have its story told. But the twentieth-century opposition to surface coal mining, particularly the movement to outlaw the practice, also has a good deal of historical significance in its own right. The campaign to abolish stripping was primarily a movement of farmers and working people of various sorts, originating at the local level, and writing its history brings attention to the role played by common folk in the conservation, preservation, and environmental movements. Additionally, many of the activists engaged in the effort to ban surface coal mining understood that struggle as primarily a dispute between property owners over the legitimate use of privately held land and its resources. Examining this part of the opposition campaign points up the importance of a tradition of veneration for private property in shaping both political consciousness and social conflict in U.S. history.

First, and perhaps most important, by focusing on the efforts to outlaw stripping this book contributes to the recovery of the environmentalism of common people. It delineates a history in which abuse of the land and its resources by coal operators generated social consciousness and activism that was somewhat independent of and distinct from the environmentalism of middle-class suburbanites and upper-class urban dwellers. The rural Appalachian farmers, workers, and unemployed who organized to ban strip mining, or to enact and enforce more stringent regulations, were typically not affluent or college-trained, and many of them were committed to exposing the linkages between the stripping that was ruining the land and the poverty that was devastating the people of the region. Like their middle-class counterparts, these critics expressed dislike for stripping in aesthetic terms, as a concern for the conservation of valuable mineral and timber resources, and as a matter of preserving the ecological integrity of the hills. But they were more likely to bemoan the damage done by strip mining to farmland and homesteads, as well as the loss of jobs in an economically depressed region. In the mid-1960s, some opponents of surface mining also began to abandon lobbying, petitioning, and working through the courts when those formal channels of resolving grievances proved ineffectual. In their place they substituted other tactics, shooting at mine employees, sabotaging mine machinery with explosives, illegally occupying strip-mine sites, and blocking haul roads. These tactics further distinguished the environmentalism of local groups of common people from that of law-abiding national environmental organizations.[3]

Yet, however militant and subversive proponents of a ban on stripping appeared to be, their efforts were neither seditious nor revolutionary in intent. Elements of class conflict and demands for social equity were present in the struggle between coal operators and "abolitionists," but both sides generally maintained a reverence for the institution of private property, albeit property on a different scale. Even when they violated the property rights of coal companies by occupying mine sites or destroying mine machinery, activists did so in the interest of protecting their own proprietary interests. In this way, the history of opposition to strip mining speaks to a long-running debate among historians about American political traditions, a debate that has revolved primarily around the question of common values.

In the 1950s, historians associated with a "consensus" school of thought downplayed the existence of significant conflict in American politics and emphasized the hegemony of liberal values, including the sanctity of private property.[4] Other scholars have since restored social divisions and conflict to the historical narrative but acknowledge an identifiable reverence for private property. Rather than locate the origins of this reverence in Lockean liberalism, however, these "new" historians find them in a republican ideology, which put a premium on propertied independence.[5]

This history of opposition to strip mining draws on the work of both consensus and new historians and suggests that at least by the twentieth century, many Americans had blended the two traditions in their values and actions. Opponents' nonviolent civil disobedience and calculated acts of violence fit easily within a Lockean doctrine of natural rights that deemed rebellion a right and duty when other natural rights, such as the right to property, were being threatened. Their concern for small-scale, private property and insistence on its importance as a foundation for independence and self-reliance also demonstrated the shaping force of a republican ideology.

Organized around these themes—the prominent role of common people in the campaign to outlaw surface coal mining and the importance of a tradition of private property in shaping the consciousness and actions of the opposition—this book is both comparative and comprehensive. It focuses on what was shared by opponents throughout the region as well as the differences among them, encompassing the strip coalfields of seven states, including middle and western Pennsylvania, eastern Ohio, West Virginia, southwestern Virginia, eastern Kentucky, eastern Tennessee, and northeastern Alabama. By concentrating on surface coal mining in Appalachia, this history slights the stripping taking place in the West and it does not investigate opposition in that region. It was not until the mid-1970s, however, around the time a federal regulatory bill was signed into law and the movement for a ban dissipated, that a considerable amount of strip mining was done in states like Wyoming, Montana, and Arizona. It was surface mining in the Appalachian coalfields that prompted the first state action, spawned abolitionist sentiment, and sparked a campaign for federal prohibition of the practice.

Chapter 1 prepares the reader for the more focused examination of opposition to surface mining that follows. It consists of brief natural and social histories of Appalachia up to the early twentieth century as well as a history of the strip mining industry. The natural history surveys the geological, biological, and chemical processes that created mountainous terrain and placed seams of coal in steep and rolling hills throughout the region. The social history begins with the changes brought to Appalachia by the encroachment of white settlers, particularly after the Revolutionary War, and ends with the dispossession of the land and its resources by land company agents. The industrial history provides basic explanations of area strip mining, contour strip mining, and auger mining as part of an outline of the origins and development of stripping, from the pick-and-axe days of the mid-nineteenth century to the more technologically sophisticated operations of the 1960s.

The second and third chapters trace the beginnings of popular opposition to strip mining through case studies of the passage of early state regulatory legislation in Ohio and Pennsylvania. The Ohio case investigates the efforts of farmers and their rural allies to enact the first controls on strip mining in that state in the late 1940s. The Pennsylvania case highlights the role played by sportsmen and organized labor, including the United Mine Workers of America, in passing amendments to the state's strip mine law in the early 1960s. Although support for a complete ban was not very evident in either of the state campaigns, their histories illuminate some of the arguments and tactics used later by advocates of outlawing stripping. Both case studies also follow the development of state-level opposition into the latter part of the 1960s and early 1970s.

The fourth, fifth, and sixth chapters investigate continued efforts to pass and enforce regulations on strip mining at the state level as well as the rise of a militant grassroots movement working for a ban. A case study in the fourth and fifth chapters locates the roots of the abolition movement in eastern Kentucky and outlines the movement's development up to a failed attempt to outlaw stripping by state legislation in 1972. The sixth chapter, covering West Virginia, also provides an overview of a failed effort to pass a state abolition bill. Both case studies detail a process of disillusionment with courts and state legislatures. Together they establish a framework for examining the shift of abolition efforts to the federal level in the late 1960s and 1970s.

Chapters 7 and 8 follow the campaign for a ban all the way to enactment of the Surface Mining Control and Reclamation Act of 1977. This part of the book continues to investigate local organizing but focuses on regional and national efforts to convince Congress to pass an abolition bill. A ninth chapter evaluates the post-1977 transformation of the opposition, when most opponents accepted a role in the enforcement of the federal surface mining law. This realignment is explored through a survey of the efforts of Save Our Cumberland Mountains, a Tennessee group, as well as an examination of Kentucky activists' successful campaign to outlaw the broad form deed. A conclusion assesses the historical traditions and cultural factors precipitating advocacy and activism for a ban on surface coal mining as well as the political maneuvering that eventually undermined the abolition movement.

chapter one

Making, Taking, and Stripping the Land

The Appalachian Mountains derive their name from the Apalachee, a group of North American aboriginal people who once inhabited present-day northern Florida and southern Georgia. European explorers of the sixteenth century first applied an altered name of the tribe to the highlands as they made their way across the southeastern part of what is now the United States. Only well after the Civil War did "Appalachia" refer to more than a physiographic mountain system. Those who studied and wrote about the region in the antebellum period thought of it solely as a place characterized by a particular topography and lithology. This understanding of Appalachia has been complicated somewhat by late-nineteenth- and twentieth-century efforts to define the area culturally and economically. But contemporary geographers and geologists alike continue to think about the region in terms of its notable surface features and rocks.[1]

Viewed as the eroded remnants of an ancient mountain system, Appalachia stretches from Newfoundland, in easternmost Canada, to northern Alabama, in the southeastern United States, a distance of nearly 2,000 miles. Yet the Appalachians are not a single range of mountains. They can be divided up into northern and southern segments, roughly corresponding to the glaciated and unglaciated sections, meeting at the Hudson and Mohawk Valleys in present-day New York. Furthermore, while the northern segment of the mountain system is undifferentiated, the southern section can be broken up into four belts or provinces, identifiable mountain groups running

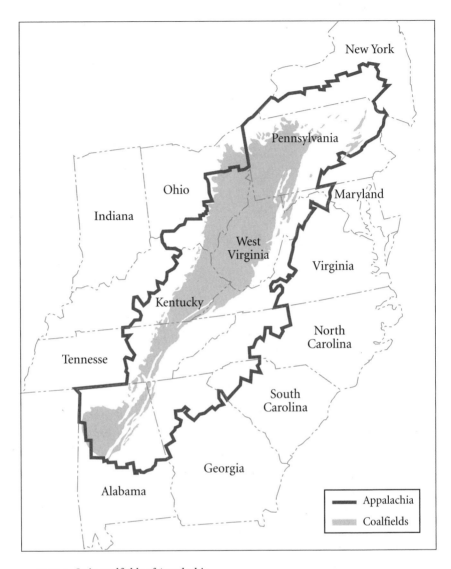

MAP 1. Strip coalfields of Appalachia

parallel to the whole chain. Each of these provinces—the Appalachian Plateau, Valley and Ridge Province, Blue Ridge Province, and the Piedmont Fold
and Thrust Belt—have distinct geological histories and appearances all their
own.[2]

Farthest northwest in the southern segment is the Appalachian Plateau,
extending from northern Alabama to eastern New York. Despite the name,
only parts of this province are true plateau, an elevated and level expanse of

land. Most of it is dissected by deep valleys and some of the area is mountainous. The province includes the Catskill Mountains in southeastern New York, the Pocono Plateau of northeastern Pennsylvania, the Allegheny Mountains from north-central Pennsylvania to southeastern Virginia, southeastern Kentucky, and southern Tennessee, as well as the Cumberland Plateau from central Tennessee to northern Alabama. The rock layers in these areas are primarily flat-lying or nearly flat-lying and sedimentary in origin, formed by the cementation of pieces of preexisting rocks or precipitated from water. They date from the late-Proterozoic to Paleozoic Eras, and much of the bituminous coal mined in Appalachia comes from the province's strata of the Pennsylvanian Period (ca. 320–286 million years ago).[3]

The second belt, moving southeastward, is the Valley and Ridge Province. This belt follows the length of the Appalachian Plateau and is characterized by narrow ridges running parallel to one another for tens or hundreds of miles. Between the ridges are parallel valleys, some of which are quite broad. The southeastern side of the belt, in fact, is nearly a single valley, varying between 2 and 40 miles across and running from New York to Alabama. Like the Appalachian Plateau, most of the rocks in these ridges and valleys are a varied succession of sedimentary rocks dating from the late-Proterozoic Era through the Paleozoic Era, including shale, limestone, and sandstone. These rocks do not lie flat, however, because extensive folding and faulting have taken place in the province. Yet, again like the strata of the Appalachian Plateau, parts of the Valley and Ridge belt are coal-bearing. The most important of these areas is in northeastern Pennsylvania, where four major coalfields make up the Anthracite Region.[4]

The third belt in the southern segment of the Appalachian Mountain system is the Blue Ridge Province. Shorter and narrower than the other three, it stretches from southern Pennsylvania to northern Georgia in a thin strip that widens somewhat in southwestern Virginia. To the north the province consists of a single massive ridge, 10 to 20 miles across, which then diverges into two principle ridges at Roanoke. The Unaka or Great Smoky Mountains of eastern Tennessee are the higher, massive northwestern ridge and the Blue Ridge Mountains of Virginia and North Carolina are the southeastern ridge. Adjacent to these mountains is the Piedmont Fold and Thrust Belt, the fourth province, extending from New England to the middle of Alabama, with the Fall Line as its southeastern edge. Both provinces have sedimentary and metamorphic rocks, dating back to the Cambrian Period and earlier, but neither has any significant coal seams.[5]

The formation of coal in the Appalachian Plateau and Valley and Ridge provinces during the Pennsylvanian Period, and elsewhere at different points in time, is reasonably well understood by geologists. They disagree, however, about the process of deposition. Coal forms when fossil plants are carbonized under high temperatures and pressures, removing various gaseous and liquid compounds and leaving carbon films. Put simply, coal is condensed and altered forms of organic matter, such as spores, ferns, conifers, and ancient scale trees. The amount of heat and pressure this organic matter is subjected to over time determines the rank of the coal, that is whether it remains as peat or ultimately becomes lignite, bituminous coal, or anthracite coal. Peat occurs at the earth's surface and consists of plant debris that has not been carbonized. Lignite is a soft brown coal in which carbonization has begun to take place but has not advanced very far. Bituminous coal is black, hard, and bright and contains more carbon than lignite but retains some volatile matter. This low-grade coal can form under the weight of a thick overlying sedimentary pile. Anthracite coal is even blacker, more dense, and shinier, with a higher carbon content and little volatile matter. The formation of this high-grade coal requires additional heat and pressure, such as that found inboard of tectonically active margins of continents. Coals of each rank are found in beds or seams—sometimes in a horizontal position as when they were deposited—but of varying thickness and distance from the surface.

In the 1930s, J. Marvin Weller noted that coal beds were part of a sequence of rocks, typically wedged between a grouping of sandstone, sandy shales, and underclay and another grouping of marine limestones and shales, and this sequence was repeated in the stratigraphic column. He explained these cycles of sedimentary rock, or cyclothems, as the result of repeated uplift, erosion, subsidence, and inundation by a shallow sea. Coal was formed during the periods of subsidence, which were marked by increased rainfall, the growth of lush vegetation, and the accumulation of peat in extensive swamps. But Weller's tectonic hypothesis, as it was called, required an unreasonably large number of uplifts and downwarps, and this problem prompted a search for another explanation. In 1936, Harold R. Wanless and Francis P. Shepard argued that global fluctuation in sea levels due to climate-induced waxing and waning of glaciers (glacio-eustasy) produced the "rhythmic alternations of sediment" during the Pennsylvanian Period, as well as in the Early Permian. Coal beds were formed across the North American continent, they maintained, when increased humidity associated with melting glaciers and a warmer climate created conditions favorable for the growth of vegetation

and swamps developed in the lowlands. A subsequent rise in sea level facilitated carbonization of swamp plants (by creating a low-oxygen environment) and resulted in the formation of shales and limestones. In time, vegetation decreased in the uplands and sands poured out on the piedmont,
bringing a sedimentary cycle to a close.[6]

Some geologists have now settled on a blend of the two earlier positions to
explain the origins of coal-bearing cyclothems across the ancestral North
American continent. Tectonically induced changes in sea level were predominant in the Central Appalachian Basin, they contend, and these were concurrent with climate-induced variations, which had a greater impact on the
Illinois and Kansas Basins farther west. Past proponents of the contending
hypotheses had simply been focusing on two end-member processes that occurred at the same time. According to George Klein and Jennifer Kupperman, mountain building along the eastern margin of the North American
continent caused rapid, short-term tectonic changes and cyclic variation in
sea level. With each flexural event, Klein explains in an article with Debra
Willard, basins were underfilled and marine waters transgressed on the land.
But mountains eroded and shed sediment, which filled the basins to sea level.
This produced the swampy conditions conducive to the creation of peat,
which later became coal. A cycle then ended with more uplift and the retreat
of oceans.[7]

Whatever their origins—and that is still a matter of some debate—the
sedimentary cycles responsible for coal deposition in Appalachia ceased by
the start of the Mesozoic Era (ca. 245 million years ago).[8] Yet the Appalachian
mountain system continued to undergo important geological changes. During the Permian and Triassic Periods some of the horizontal beds of rock laid
down as part of sedimentary cycles were subjected to folding and compression. In northeastern Pennsylvania this pressure and the accompanying heat
metamorphosed the coal into anthracite. Pressure drove off gases and impu-
rities, increased the proportion of carbon, and left an organic compound
with a high heat output and low ash content. On the eastern edge of the Appalachian Plateau, coal was similarly affected by folding and compression,
but it was not subjected to enough pressure to transform it into anthracite.
In areas further to the west, beyond the deformation zone, the coal retained
much of its volatile gases and sulfur.[9]

Taking the Land: Dispossession of Early
Inhabitants and Capitalist "Development"

When Europeans first stumbled upon North America in the fifteenth century there were as many as fifteen indigenous tribes living in the southern mountain region. By the eighteenth century, however, many of the tribes had been decimated by diseases, a result of the exchange of European pathogens to which they had no immunities. For those tribes that did persist, continued epidemics and other factors weakened the ability of the indigenous people to resist the encroachment of white settlers, and the Europeans took their lands. The Cherokee alone lost 40 percent of their territory by the end of the Revolutionary War, and they ceded an additional 3 million acres between 1800 and 1819. Through a campaign of so-called Indian wars, U.S. soldiers displaced (either exterminated or forced to reservations) nearly all the bands of native people between the Great Lakes and the Gulf of Mexico, opening up the northwest and southwest territories for white settlement.[10]

Yet, as Wilma Dunaway explains, not all white settlers fared equally well in Appalachia during its frontier years. Federal laws designed to protect the rights and promote the interests of aspiring homesteaders were not implemented until the middle of the nineteenth century, and these laws were designed to regulate settlement in the Midwest. During the preceding decades northeastern merchant capitalists, land companies, and southern planters managed to expropriate most of the Appalachian region's total acreage. By the mid-1700s tidewater planters and British Court favorites had acquired much of southwest Virginia, present-day West Virginia, and western Maryland. In western North Carolina, planters and two land companies monopolized a good portion of the northern sector. By the end of the eighteenth century, probably three-quarters or more of eastern Kentucky's frontier lands were held by absentee speculators. And in Tennessee, merchant capitalists, land companies, and distant planters amassed more than two-thirds of the territory's mountain region. Having engrossed the land with the purpose of making a profit, these speculators charged high prices for their acreage and, as a result, at least two-fifths of settler households were without land in the antebellum period.[11]

Due to inequitable patterns of land ownership, eighteenth- and nineteenth-century Appalachia did not initially exemplify Thomas Jefferson's vision of a democratic, egalitarian society based on independent, small freeholders. A speculative market in land gave rise to social stratification, which characterized rural communities of the region just as it did industrial cities of the

Northeast and Midwest. Yet even as the engrossment of land imper
mountain settlers' dreams of a "competency," most of them contin
lieve in and strive for the Jeffersonian ideal. Their qualified success
ing a life of propertied independence—often by squatting on land
others—is evidenced by the fact that generally self-sufficient family farms
did eventually become the backbone of the Appalachian economy. By 1880,
Appalachia contained a greater concentration of noncommercial family farms
than any other part of the nation.[12]

Through the first half of the nineteenth century, however, the inhabitants
of the southern mountains were still not "Appalachians." The creation of Appalachia as a coherent region inhabited by a homogenous population with a
uniform culture, as Henry Shapiro has put it, was a post–Civil War phenomenon. In the antebellum period, travel literature presented mountain people
as no different from other Americans, or at least as no different from other
southerners. A few accounts even incorporated an awareness of the ways in
which the southern highlands were internally differentiated, recognizing the
difficulty of making generalizations about the mountains and its inhabitants.
But the 1870s saw the rise of local color writing, which focused on the supposed peculiarities of non-urban people and places. With it came the liter-
ary, social, and economic transformation of Appalachia. Though the local
color genre was not limited to descriptions of the southern highlands, it was
there that it had the most significant and lasting impact.[13]

One of the first writers to assert the "otherness" of Appalachia was Will
Wallace Harney, who published "A Strange Land and Peculiar People" in *Lip-
pincott's Magazine* in 1873. Similar essays by other local colorists followed. In
more than two hundred travel accounts and short stories of the local color
variety published between the early 1870s and 1890, southern mountaineers
were presented as backward and isolated from the mainstream of American
life. By the turn of the century, various individuals were citing this regression
as the basis for a mission of uplift. In an address entitled "Our Contemporary Ancestors in the Southern Mountains" (1899), Berea College president
William G. Frost called attention to the special needs of "mountain whites"
and, like many others, compared "Appalachian America" to Revolutionary
America. Both had the same population, he said, the former having pro-
gressed little beyond the latter's level of civilization. For Frost and other observers, as Allen Batteau notes, it was not so much strangeness as familiarity
that made Appalachia interesting and worthy to America. The roots of an
American cultural identity could be found in the people of the southern
mountains: to be Appalachian was to be quintessentially American. Yet the

perception of mountaineers as an isolated people of another time also required ameliorative programs of action.[14]

The social construction of an Appalachian people spoke to the need for a rising, urban-industrial middle class to see southern mountain people as a repository of an increasingly threatened republican inheritance. Yet to missionaries, entrepreneurs, and others shaped by late-nineteenth-century urban-industrial transformation, southern highlanders also badly needed modernization. To missionaries, the mountaineers whom they romanticized as "contemporary ancestors" needed social uplift, the betterment that would bring their values and mores up to date. They were deserving of help because of the independence and individualism fostered by geographic isolation, but their feuding, moonshining, and parochialism had to be addressed if Appalachia was not to remain a pocket of backwardness. "[W]here the local colorists had been content to see mountain life as quaint and picturesque, and for this reason inherently interesting," Shapiro contends, "the agents of denominational benevolence necessarily saw Appalachian otherness as an undesirable condition and viewed the 'peculiarities' of mountain life as social problems in need of remedial action."[15]

Capitalists, on the other hand, interpreted modernization to mean development of the region's resources. Such development came to the mountains under the rubric of a New South Creed, a post–Civil War ideology emphasizing diversified agriculture, industrialization, and urban growth. The Creed was promulgated by native private speculators who had taken stock of the South's timber and mineral resources, as well as its proven agricultural potential, and sought to entice railroads and northern investors. In some respects the speculators were successful, in the South as a whole and the mountains in particular. By 1900, four major railroad lines had entered the southern Appalachians and numerous branch lines extended from these main lines. This increase in track mileage, which provided new links between isolated hollows and the cosmopolitan world beyond, brought dramatic changes to the region. But the changes that followed diverged from the vision of a modern South evoked by proponents of the New South Creed. In Appalachia, logging camps and coal towns proliferated, agriculture declined, and the one-way flow of resources out of the region left its people impoverished. Increasingly, the southern mountain economy developed on the periphery of a distant core, its resources owned by outside investors and removed to fuel industrial expansion in the North.[16]

The rise of the new extractive industries hinged on acquisition of land from resident landholders. This dispossession was sometimes lawful, a legitimate exchange of surface or mineral rights for cash. Such transfers were often facilitated by local entrepreneurs who knew the land, its occupants, and what it would take to get them to loosen their hold on the property. And there were, in fact, many reasons why property owners would want to sell part or all of their land. High birth rates, population growth, and land scarcity had made subsistence farming increasingly difficult for each new generation, and mountain life was never idyllic. Travelers' accounts, census returns, government reports, and demographic studies all indicate that after 1830 population increases began to exert pressure upon available economic resources. At the same time, new stock laws required animals, rather than crops, to be fenced. By bringing an end to open foraging the laws increased the cost of raising livestock and limited the opportunities for mountaineers to rely on the animals for food or infrequent commercial exchange.[17]

The many factors making subsistence farming increasingly difficult with each new generation meant that the first forays of land agents into the region usually were welcomed. But the agents did not receive a kind reception from everyone and, over time, an increasing number of landowners were reluctant to sell land coveted by timber and mineral companies. In these cases, agents acquired surface and mineral rights through illicit methods. Ownership of property in the region was often uncertain because of confusion surrounding the original grants and the subsequent purchase of the land by other settlers or occupation by squatters. Land titles were obscure, deeds were lost, and records were poor in most mountain counties. Speculators with a better understanding of laws, courts, and the workings of local and state governments used their knowledge and connections to their own advantage. As a result, by 1910 outlanders controlled not only the best stands of hardwood timber and the thickest seams of coal but a large percentage of the surface land in the region as well.[18]

For many mountain residents, partial or complete land dispossession brought an end to their reliance on farming to meet basic needs and the beginning of a new livelihood cutting timber. Southern highlanders also joined the wage labor force digging coal as operators opened underground mines across the southern highlands. Beginning in the 1880s, small companies dedicated to industrial mining proliferated as branch lines extended off main railroad lines, providing access to national markets. In the 1890s, bigger mining companies came to the mountains just as the market for Appalachian coal expanded. The region offered cheaper freight rates and lower labor costs than the northern Central Competitive Field, which had been organized by the United Mine Workers of America in 1898, and the new industries of the

Northeast and Midwest shifted their coal purchases accordingly. Dramatic increases in mining followed. West Virginia produced nearly 5 million tons of coal in 1887, but this had risen substantially to 90 million tons by 1917. In Appalachia as a whole, coal production increased fivefold between 1900 and 1930, eventually accounting for nearly 80 percent of national production.[19]

Most of the first generation of Appalachian miners worked in the mines only seasonally, taking time off to plant, grow, and harvest crops on the land remaining to them. Even after being absorbed completely into coal camps, mining families continued to keep gardens for their subsistence needs, a practice encouraged by operators through their provision of free fencing and plowing as well as seeds and fertilizer at cost. Yet the livelihood of many mountaineers changed substantially with the expansion in mining, and the population of the region changed as well. Because local people could not adequately supply the mines with a labor force, coal operators imported African American migrants from the South and immigrants from southern and eastern Europe. In 1880 there were very few black miners in the state of West Virginia, but by 1910 there were 12,000. During the same period, the number of European immigrant miners in the state rose from 924 to 28,000, many from southern Italy. Through the early twentieth century, as the population swelled and diversified, the Appalachian region had one of the highest population densities in rural America.[20]

In many of the areas near logging operations and mines, civic development followed the resettlement of older residents and the settlement of new populations. Yet most of the people of the region did not profit from the extractive industries. This is the enduring paradox of Appalachia, that the inhabitants of a land so rich in natural resources could be so poor. When earlier generations of scholars tried to explain this problem they missed the significance of the economic activity in the late nineteenth and early twentieth centuries. The poverty of Appalachia, they claimed, was due to a lack of modernization. Others tried to improve on this explanation by suggesting that a culture of poverty held back mountaineers from taking advantage of the opportunities offered by urban and industrial development. Ronald Eller and others have argued, however, that Appalachia is marked by poverty not for a lack of modernization or because of inhibiting cultural traits, but as a consequence of a particular type of modernization. Between 1880 and 1930, they explain, the region supplied the raw materials essential to the factory and mill production in the Northeast and Midwest. Because so much of the timber, minerals, and land had been bought up by northern speculators, and many of the companies were controlled from outside the region, the great wealth of the land flowed out from the highlands never to return. As a result, many of the people and much of the environment of Appalachia have been impoverished.[21]

Stripping the Land: From Picks and Shovels to Draglines

The coal mines opened by operators in the late nineteenth and early twentieth centuries, in Appalachia and other parts of the United States, were not all deep mines. An increasing number were one or another type of surface mine. As practiced in the early twentieth century, surface coal mining included area and contour mining, both of which involved removing an overlayer of rocks and soil to get at a coal seam. In the years after World War II surface coal mining also encompassed auger mining, boring into an outcropping of coal with a huge auger. And by the mid-1970s, some strip operators were extracting coal by a process vividly referred to as "mountaintop removal." In the debate over surface mining—in hearings, protests, and literature on the subject—opponents and coal industry officials alike often failed to distinguish between these different methods. Instead they made reference to "strip mining," "stripping," or sometimes "surface mining." Usually they meant contour mining, the most common form of surface coal mining in Appalachia. Sometimes they were also talking about auger mining, which eventually became a normal part of most contour operations. In other instances, particularly if they were from eastern Ohio, their concern was area mining. This lack of specificity is understandable, and at times in this book I conflate the methods for the sake of convenience. But the differences between the various types of strip mining are important for understanding the rise of a concerted abolition effort. The methods had distinct histories, shaped largely by evolving technology and changing market conditions, and they did not impact the land or the people in exactly the same ways.

Surface coal mining probably began in North America during the colonial period. Except for scattered reports of coal in present-day Illinois between 1660 and 1680, coal was first found in what became the United States underlying the James River, in Virginia, just after the turn of the eighteenth century. One of the earliest references to a form of surface extraction refers to this coalfield and was made by Dr. Johann D. Schoepf upon his visit to Richmond, Virginia, in the winter of 1783. Twelve miles outside of the city, south of the James River, the wind had blown a tree over, exposing white clay-slate, a black clay-slate, and a coal bed. This afforded an opportunity for people to mine the sulfur-laden coal, which they sold at the river for one shilling per bushel. Such early surface mining was typically done by farmers and common town dwellers for local exchange and use. Yet the pick and shovel, which were the miners only tools, limited their efficiency and destructiveness. Coal was mined where it could be seen, and removal of any part of the surface over-layer by miners themselves was minimal. In most cases, some type of natural perturbation or weathering exposed parts of a bed and very little further excavation was necessary to bring the mineral out or up.[22]

During the nineteenth century, there was an increasing interest in making a business out of strip mining by adopting new methods and expanding production. Coal provided an alternative fuel source to dwindling supplies of wood, and that meant there was money to be made in its extraction. By the 1820s, mountain residents in eastern Kentucky were being hired by entrepreneurs to dig coal from shallow beds to supply growing markets in Lexington, Frankfort, and Louisville. Like their counterparts elsewhere, the eastern Kentucky miners first exploited visible coal seams, such as along eroded stream banks. By the mid-nineteenth century, however, the miners were using steel scrapers, drills, and black powder to expose coal beds. They hitched horses or mules to the scrapers and alternately plowed and removed "overburden," the soil and rocks above a coal seam. Once they had removed the surface soil, a "shaker" sat with a churn drill between his knees while a "driver" administered a blow to it with a sledgehammer. After each strike of the hammer the shaker lifted the drill, gave it a half-turn, and thus gradually sunk it into the cap rock to the coal seam. The holes created by repeated drilling were tamped with black powder, set off by a slow-burning fuse, and the loosened rock was pushed aside. The miners then shoveled the coal onto wagons that transported the mineral to rafts on nearby waterways.[23]

Farther west, in the Danville region of Illinois, miners also used steel scrapers to work the local coal beds. In 1866, Kirkland, Blankeney and Graves opened the first Illinois strip pit on Grape Creek and this was followed, in 1875, by the establishment of a surface mine in Hungry Hollow, owned by Michael Kelley. Although similar to what was being done in the Appalachian coalfields at the time, the mining in the Midwest more closely approximated what became known as area stripping. Area strippers worked coal beds over a period of years, with new cuts made by alternate plowing and scraping in parallel strips. The overburden from the first cut was set to the side in an elongated mound, but the mine waste generated with each new parallel strip was placed in the adjacent pit of the previous cut. Initially, all of the plowing, scraping, and coal hauling relied on teams of horses, but in the late-nineteenth century the efficiency of area strip mining was greatly improved by application of steam technology. The first recorded use of a steam shovel at a surface mine was in 1877, at a Pittsburgh, Kansas, operation owned by J. N. Hodges and A. J. Armil. Illinois operators readily adopted "mechanical" area strip mining and, by the turn of the century, miners there were removing overburden and digging coal with steam-powered machinery on a more extensive scale than in any other state.[24]

Mechanized surface mining also moved into the coalfields of Indiana. The state's coal seams were relatively thick, averaging 3 to 5 feet in large areas and 6 to 8 feet over lesser areas, and they were close enough to the surface to make area stripping quite feasible. With the adoption of steam technology, strip production in Indiana increased nearly thirteenfold between 1914 and 1936. In the 1920s, Oakland City, Indiana, became home to what was probably the largest strip operation in the world. Run by the Enos Coal Mining Company, the mine produced 1 million tons a year. As large and advanced as it was, however, the operation still employed some primitive surface mining methods, including team-drawn scrapers, picks, shovels, and wire brooms to remove dirt from the top of the seam. In fact, the scraper method was not replaced at U.S. mines until after 1936, when the Traux-Traer Coal Company first substituted tractors for its horse-drawn implements at one of its operations in Elkville, Illinois. But area mining did become more sophisticated and expand in the interwar years in Indiana. By the latter part of the 1930s, it had more strip pits than any other state, producing 8.2 million tons of coal, making it the leading producer behind Illinois. As a percentage of total coal production, stripping in the state also jumped dramatically from 4.6 percent in 1920 to 53.2 percent in 1940, taking the lead over underground mining.[25]

In Ohio, as in Indiana, area stripping with steam technology began just before World War I, when the nation was poised for another coal boom.[26] One of the state's first surface operations opened at Rush Run, on the Ohio River, in 1914. In its first two years only 25 acres of a 200-acre tract had been mined there, but Ohio strip coal operators quickly opened other mines and continually introduced the latest technological innovations. The Apex Coal Company started a strip operation in Harrison County, the Kehota Mining Company opened another in Perry County, and sometime before 1916, the Piney Creek Coal Company introduced an electric-powered steam shovel at its strip operations near Steubenville, in Jefferson County. In the 1920s, Ohio coalfields also witnessed the introduction of the first draglines, which maneuvered large buckets by steel cables rather than a fixed boom. Combined with improvements in the methods of transporting coal, electric shovels and draglines greatly facilitated increased production in the state. By 1938, Ohio's seventy-two strip mines produced 2.5 million tons of coal, making it the fourth leading strip producer in the country, behind Illinois, Indiana, and Missouri (where strip mining production quickly declined). Expanded production during World War II increased output even more dramatically to 17.3 million tons, and strip mining edged closer toward eclipsing deep mining. Between 1926 and 1947 the percentage of coal mined by stripping in the state jumped from 9 percent to 46 percent.[27]

Although much of the industry's growth was occurring in the Midwest, during the late-nineteenth and twentieth centuries surface coal mining also evolved and expanded in the mountainous parts of Appalachia. Operators in Pennsylvania's anthracite fields began to use steam technology as early as 1881, but they worked coal seams by what was referred to as contour strip mining. In contour stripping, miners created a "bench" on the side of a mountain where coal was exposed or near the surface. Strippers scraped away overburden along a ridge top and constructed two surfaces, one that was vertical, called the highwall, and another at the level of the coal seam that was horizontal and met the highwall at its base. The L-shaped bench created by the intersection of these two surfaces extended a variable length, linked at intervals to access roads, and sometimes wrapped back around a ridge. The overburden removed in the process was pushed down the mountain slope, creating a "spoil" pile, or, less often, it was stacked on the bench. As miners discarded the rock and soil over-layer they also extracted the coal below it, widening the bench in the process, until the overburden was too thick to make the operation economically feasible. By the early twentieth century up to 10 feet of overburden could be removed profitably for each foot of coal in the seam.[28]

Difficult terrain and primitive equipment initially hampered the expansion of contour surface mining in Pennsylvania, as was the case in other Appalachian states, but production rose steadily after World War I, and soared during World War II. The process of mechanization and the industry's expansion in the state can be seen in the increase of the number of power shovels at anthracite and bituminous strip mines. Through the 1920s there were less than 100 power shovels working the eastern anthracite operations, but by 1936 there were 364. That number nearly doubled by 1947, to 609, and almost half of these shovels were draglines. In the bituminous fields of western Pennsylvania, the number of power shovels also increased between 1936 and 1947, from a minuscule thirty to more than a thousand, the great majority of which were diesel and gasoline dragline excavators. Such a dramatic mechanization of Pennsylvania mines contributed to a sharp increase in production. In 1915, the eastern anthracite region produced 1.2 million tons of strip coal, or 2.5 percent of all anthracite mined in the state. By 1947, production had increased tenfold to 12.6 million tons, nearly a quarter of all anthracite coal mined in Pennsylvania. As late as the 1930s, western bituminous strip coal operators mined only 750,000 tons, but this jumped to 37 million tons in 1947, declined to 19 million tons in 1958, and slowly edged toward 30 million tons in the 1960s.[29]

Mechanized contour strip mining also spread to other Appalachian states and slowly expanded during the years after World War II. In eastern Kentucky, the first strip mine employing steam power shovels opened in 1905, when the Lily-Jellico Coal Company contracted with the Robinson Creek Construction Company to surface mine a seam at Lily, in Laurel County. These first shovels were Vulcan railroad-type shovels, with buckets of 1 cubic yard capacity and mounted on railroad tracks. They worked in pairs, with one to strip the overburden and the other to load the coal, and the mineral was hauled out by horse-drawn wagons to a tipple, where it was loaded for shipment by rail. But the Lily strip mine was one of only a few in eastern Kentucky. In 1945, there were seven surface operations in the state, with a combined production of 130,000 tons of coal. By 1947, the number of mountainside surface operations increased to thirty-eight, still relatively few compared to Pennsylvania, and production was 1.9 million tons. Nearly a decade later, the area had seventy-two strip pits, which actually produced 8,000 tons less than the lesser number of 1947 mines. Not until 1958, when the number of mechanized contour operations increased to ninety-five, did production top 2 million tons.[30]

In West Virginia, contour strip mining with steam technology began in 1916. Fuel needs during World War I brought a brief expansion of the industry there, but the growth was only temporary. Maximum production in the state in the early 1920s was a scant 296,000 tons. By 1938, only the northern Brooke and Hancock Counties reported strip mine production, with a combined total of 207,000 tons for that year. Taking advantage of more powerful equipment and responding to nearly unlimited demand for coal during World War II, however, West Virginia strippers greatly expanded their operations. Production increased tenfold between 1939 and 1943, with much of the early new activity concentrated in the northern part of the state, where the coal was thick, the overburden was well adapted to stripping, and hard surface roads provided easy access and mineral transport. The center of this mining boom was in Harrison County, where production skyrocketed from

50,000 tons in 1940 to 3 million tons in 1943, and accounted for nearly half of all the strip coal produced in the state during World War II. In the following decade, West Virginia surface mine production averaged 10 million tons annually, and by the mid-1960s it made up 10 percent of all coal mined in the state.[31]

In Tennessee and Virginia, mechanized contour surface mining had an even slower start, but as in other mountain states production eventually reached significant levels. Commercial contour operations in Tennessee date back to World War I, when old Panama Canal equipment was put to use in Grundy County, but these strip mines were short-lived. No production was reported for the state in 1936 and a decade later, at the end of World War II, Tennessee produced only slightly more than a half of a million tons of strip coal. Yet, by 1955, production figures had increased threefold. In 1963, Tennessee had fifty-eight strip mines, nearly all in the Cumberland area, which produced 2.5 million tons of coal. Likewise, operators in Virginia began to surface mine coal in significant amounts after mid-century. In 1947, the state had fifteen strip mines, all in the far southwestern counties, with a total production of 1.1 million tons of coal. In the 1950s the number of contour operations doubled, though production dropped below 1 million at mid-decade. Significant expansion occurred only after the national recession in 1958. In 1963, the number of contour surface mines in Virginia increased to 130 and production jumped sharply to nearly 7.5 million tons.[32]

As contour surface mining spread and expanded in Appalachian states it was modified by the advent of auger mining. At first this method was employed after contour stripping had been concluded to obtain more, but not all, of the otherwise irretrievable coal. Increased strip mining during World War II left many miles of highwall containing exposed coal, and after some experimentation operators developed large, efficient augers to recover the mineral from these seams. Augers, which were usually several feet in diameter, were driven into the foot of a highwall and coal came out of the hole like wood shavings produced by a drill bit. With additional extensions, augers could penetrate farther into a seam, sometimes reaching old tunnel-and-pillar deep mines, removing even more of the mineral. But auger holes had to be smaller in diameter than a seam was thick, and they were spaced a few inches apart. Consequently, the machines brought out little more than half the coal. The method was wasteful but relatively cheap, and operators increasingly relied on it in the 1960s to profitably strip previously unmined ridges. An auger operation working a 4- to 6-foot virgin coal seam could realize a net profit of close to a dollar per ton. A large enough auger could load

15 tons of coal in less than one minute and, if the trucks could haul it out at that rate, the total profit might amount to millions of dollars in a few years. Not surprisingly, then, production at auger mines increased rapidly from less than 2 million tons in 1952 to 7 million tons in 1958, and a total of eight states produced 12.5 million tons of coal by auger in 1963.[33]

Kentucky, West Virginia, and Ohio quickly took the lead in auger mining after its introduction. The method was first used in Kentucky in 1949, when the Blair and Oldham Coal Company opened a strip mine near Isom, in Letcher County. Within a decade, eastern Kentucky auger mines were producing nearly 4 million tons annually, or a quarter of all auger-mined coal in the United States. By the mid-1960s, the area was home to the largest coal auger in the world—with a 7-foot bit that dwarfed all earlier machines—and a total of 202 auger mines, which produced 9.5 million tons of coal. Yet West Virginia followed close behind Kentucky in terms of tonnage. In 1963, the state's auger mines produced 3.7 million tons of coal and, in 1970, they reached a production peak of 5.7 million tons. Auger mining (as well as contour mining) also expanded in southeastern Ohio. In the early 1960s, the state was ranked third in terms of total auger production, mining nearly 2 million tons of coal by that method. Other states with active auger operations by that time included Virginia and Pennsylvania, both of which produced more than 1 million tons annually, as well as Tennessee and Alabama, which mined a quarter of a million tons and 100,000 tons, respectively.[34]

Contour strip mining was modified again in the second half of the 1960s, when operators made early attempts to mine coal using a method that later became known as "mountaintop removal." This occurred almost exclusively in Kentucky and West Virginia. In areas where coal seams were close to the top of a ridge, some strippers dispensed with following its contours and took off the whole top of the hill instead. Miners blasted and scraped away the soil and rock overburden and pushed it over one or the other side of the ridge until the coal was exposed. This could decrease the altitude of a hill by as much as 20 percent, while simultaneously increasing its thickness. But the machinery available to operators using this method was nearly the same as what strippers used on contour operations. This limited the amount of overburden that could be moved and greatly restricted the possibilities for mountaintop removal until technological improvements in the early 1980s.[35]

From the beginning of the twentieth century to its end, in fact, it was technological innovation that facilitated the steady growth of surface coal mining. Like other sectors of the economy, the industry responded to the expansion and contraction of demand, particularly the dramatic increase in use of coal by electric power utilities. But larger and more powerful earth-moving equipment made it possible for strip operators to respond to this demand and take market share from deep mines. In 1936, there were still ninety-six "horse stripping operations." By the late 1950s, horses were gone from the mines and nearly all fixed-boom steam shovels had been replaced by diesel- and electric-powered shovels and draglines. The majority of these still had low dipper or bucket capacity (around 3 cubic yards), but new shovels and dragline excavators were capable of moving ever-larger quantities of over-burden and coal. Between 1945 and 1963 the number of dippers or buckets with a 12-cubic-yard capacity or greater more than doubled, from 75 to 154. In 1956, the Hanna Division of Consolidation Coal introduced the first of a number of "monster" draglines. Their twelve-story excavator, dubbed the "Mountaineer," dug for coal 90 feet below the surface near Cadiz, Ohio. And at the other end of the surface mining production process, the movement of coal from mine to tipple was facilitated by the steady increase in truck size. At the close of World War II, the average capacity of trucks working strip operations in the United States was 9.4 tons. By 1958, Pennsylvania had more than 1,800 trucks working strip operations, the greatest number of any state, with an average capacity of 11.4 tons. Kentucky had nearly 500 trucks, with an average capacity of 14.4 tons.[36]

With greatly improved technology, surface coal mining increased and strip operations produced nearly one-third of the bituminous coal mined in the United States in 1963. Expansion occurred most dramatically in Ohio, Penn-sylvania, and Kentucky, the leading surface mining states in the country, each producing more than 20 million tons of coal by contour and auger mining in the early 1960s. Larger shovels and bigger trucks also meant that the produc-tivity of surface mining continued to rise. In 1914 stripping produced 5 tons per man-day, and underground mining was only slightly less efficient, pro-ducing nearly 5 tons per man-day. By 1958 stripping could produce 21.5 tons per man-day as compared to underground mining's 9 tons. It was this rela-tive efficiency, in addition to the ease with which coal operators could dis-place environmental and other social costs of surface mining onto the pub-lic, that made surface mining so attractive to coal companies. And it was the displacement of the environmental and social costs of surface mining that generated early demands for regulatory legislation to control stripping.[37]

chapter two

Our Country Would Be Better Fit for Farming

Opposition to Surface Coal Mining in Ohio

In the late-eighteenth and nineteenth centuries, more than a few American farmers took up surface coal mining as a secondary occupation. During the winter months, between the harvest and planting seasons, they worked local seams to provide fuel for their families or to sell in area markets. This strip mining, which was done on a small scale with picks, shovels, and mule-drawn scrapers, fit easily within the patterns of daily life and had only minimal impact on the land. In the twentieth century, however, as the surface mining industry mechanized and expanded, stripping increasingly came into conflict with agriculture as a competing land use. This generated opposition *within* farming communities and led to some of the first state regulatory legislation.

On the eve of World War I, farmers, coal miners, and local businessmen in Boonville, Indiana, began a campaign to stop strip mining by passage of a state law. They objected to stripping because it made hundreds of acres of "fine farming land" unfit for cultivation, threw deep miners out of work, and undermined the tax base. In the 1930s opposition spread beyond Boonville and, in 1941, Indiana became the second state (after West Virginia) to enact a control law. This legislation was meant to head off growing support for a ban by requiring "the conservation and improvement" of lands that had been strip-mined, setting up permitting and bonding procedures, establishing penalties for violations of the law, and prohibiting classification of stripped areas as forest lands for taxation purposes.[1]

Farmers also played an important role in the early opposition to surface coal mining in Ohio, the subject of this chapter. Between World War I and the end of World War II, strip mining caused extensive social and environmental devastation in the eastern part of the state, ranging from massive out-migration of local populations to acid mine drainage and sheet and gully erosion. Responding to these adverse effects, farmers and other rural people organized and lobbied for passage of the first state controls on stripping in 1947. Their argument for regulatory legislation was threefold. They claimed that coal surface mining threatened agricultural productivity and the communities that farming sustained, violated God's injunction to be stewards of the land, and destroyed the aesthetic virtues of a pastoral scene. In the 1950s and 1960s, when the first legislation proved to be inadequate, opponents of strip mining used similar arguments to pass stricter control bills and improve enforcement by the state regulatory agency. By the 1970s, however, farmers and their concerns were no longer predominant in the Ohio opposition, and the focus of opponents shifted from state to federal action.[2]

Early Impact and House Bill 314

As surface coal mining expanded during the first half of the twentieth century, it had a significant impact on the land, water, and human communities of eastern Ohio. Between 1921 and 1945, coal stripping in twenty-two Ohio counties directly affected at least 22,750 acres and indirectly impacted another 5,000 acres. In the southeastern part of the state, where the topography ranges from hilly to steep, the surface-mined land had little agricultural value, selling for as low as $4 to $8 an acre. But the land in eastern Ohio proper was highly developed farm and dairy land, especially in the wide fertile valleys of the river basins, with an average value of $100 an acre. Much of the strip mining in the state was in this area. In Harrison, Jefferson, Columbiana, and Tuscarawas Counties, unregulated strip mining ruined land for future crop production and grazing, and early on, some strippers admitted as much. "[T]he coal stripping absolutely destroys the land for farming purposes," explained an anonymous author for a 1916 issue of the trade journal *Coal Age*. The ground was left rough and the overburden of soil, shale, clays, and limestone stacked bottom-up in high ridges, making it "hard to imagine what further use could be made of such land."[3]

In addition to creating a rippled landscape and reshuffling the soil layers,

MAP 2. Counties of eastern Ohio

early twentieth-century strip mining in Ohio exposed acid silt shales and in-creased erosion. As a result, nearby streams were contaminated by acid mine drainage as well as silt, killing off aquatic life. Acid contamination of ground-water also threatened local water supplies. Several small communities (in-cluding Cadiz, Wellsville, Sugar Creek and Smithfield) took their municipal water supply from wells close to stripping operations and some were forced to implement expensive filtration and chemical treatment. Where acid drain-age was acute, there was fear that communities would eventually be forced to import treated surface water, which would have been a considerable expense. In a few southeastern counties calcareous shales and limestone in the over-burden neutralized the acid. But these spoils often had their own unique problem, containing clay and clay-forming shales that made for tight, impervious, poorly aerated soil after disintegration and settling.[4]

Because it degraded farm and dairy land, strip mining also affected the tax base of counties. Once mining was completed, operators frequently abandoned mine sites and defaulted on their taxes. In other cases, depreciated property values reduced revenues. There were large increases in valuation when strip mining began, with new valuations ranging from $60 to $300 an acre. In tax dollars this could translate into an increase of more than $2 per acre. But after a year or two, with the mining finished, county auditors devalued the land to between $5 and $15 an acre, and tax revenues declined precipitously to as low as five cents per acre. In Tuscarawas County none of the land that had been stripped up to 1947 was in tax delinquency, but between 1915 and 1940 coal-stripped land there suffered a 94 percent decline in assessed valuation. This compared with a 22 percent decline in the valuation of unstripped agricultural land.[5]

A survey of a strip-mined farm in Tuscarawas County, drawn from a set of four case studies done by Charles Victor Riley in the mid-1940s, is illustrative of the damage done as a whole to the land and its inhabitants. The land Riley labeled "Unit II" had rolling to hilly topography and in 1915 was a farm of 168 acres, with three orchards and two grazed woodlots of approximately 14 acres. At the time, the land was assessed at $7,290, or $43 per acre, but three years later the farm was sold to Wayne Coal Company for $16,800. By the late 1940s Unit II was classified as abandoned farmland, with 55 of the original 168 acres stripped or affected by mining operations. Full-time farming had been discontinued in 1918 and the fields were covered with poverty grass, goldenrod, cinquefoil, broom sedge, aster, sheep sorrel, and other opportunistic plants. All the fields suffered from severe sheet and gully erosion, with gullies varying from 3 to 15 feet deep and from 6 to 16 feet wide at the top. Soil samples indicated a pH ranging between 4.7 and 5.9 (acidic), and available phosphorous and potassium were low. Of the two small streams crossing the unit, one was polluted by mine waste. In this condition, the assessed value of the land in 1940 was $1,059.46, or $10.29 per acre. In 1946, the gross income from the property was $115, most of which came from an oil lease. Typical of stripped land throughout eastern Ohio, Unit II was practically worthless after coal operators had finished with it.[6]

Such a devastating impact on farmland raises a question about why property owners would lease or sell to surface coal mining companies. Operators maintained that farms were being abandoned, coincidentally, at the same time they developed an interest in mining. This was probably true to some extent. Between 1935 and 1945, Harrison County alone lost 420 farms, and the number of acres farmed declined from 215,000 to 195,000 during the war.[7] Not all of these and other losses of farm and dairy production in Ohio were due entirely to stripping, yet surface mining in a local area was often an important factor in a farmer's decision to sell off crop or grazing land. Residents frequently banded together to resist selling out to strippers, but once isolated individuals began to sell, the process could snowball until an area was nearly depopulated. In other cases, farmers were hard-pressed just coming out of the Depression or were simply enticed by the money offered by strip mine operators. They were reluctant to insist on reclamation procedures, however, for fear the coal company would not buy the land or, in the case of a lease, would lower the royalty payments. As a result, without a state law stipulating reclamation procedures, between one-third and one-half of the lands surface mined before 1947 were not reclaimed in any way. When operators did reclaim they often did little more than attempt typically unsuccessful reforestation, without leveling the spoil banks.[8]

Throughout the 1920s and 1930s, reclamation of surface mined lands was dealt with primarily by state agricultural experiment stations, while the problem of acid mine drainage was addressed largely by engineering experiment stations and the U.S. Public Health Service. These agencies carried on their own research and funded other studies, investigating the possibilities for establishing forests, creating wildlife havens, and constructing recreational lakes at former strip sites. In 1937, the U.S. Forest Service established the Central States Forest Experiment Station, which quickly took the primary role in research on reforestation of stripped lands. With its headquarters in Columbus, most of the work of the agency was concentrated in Ohio until 1946, when it began conducting studies in other states in the central region as well, particularly Indiana and Illinois. Also, in 1940, six large coal companies with operations in Ohio formed the Ohio Reclamation Committee to facilitate their participation in government research as well as to provide a cooperative forest planting service to its members. The organization was renamed the Ohio Reclamation Association in 1945, when its combined membership of thirty-one stripping companies performed 65 percent of the surface coal mining in the state. By 1947, the association had planted 1.8 million trees on spoil banks and surrounding areas. Yet none of these spoil banks were leveled before planting and there was little effort to restore the land for crop farming or even grazing.[9]

Responding to the lack of reclamation in some areas and its inadequacy in many others, William F. Daugherty, a Democratic member of the House of

Representatives from Wellsville (Columbiana County) introduced the first strip mine control bill in the Ohio legislature in 1937. The bill mandated that future strip mine contracts for the extraction of coal as well as other minerals include provisions for "the replacement and leveling up of the earth ... so that the land is left in substantially the same condition after the completion of the mining operations as it was before the mining operations began." By exclusively regulating strip mining contracts, however, the act would not have affected operations in which the land was purchased outright. It also failed to establish administrative machinery to ensure enforcement, and some of its standards for reclamation were vague enough to be left open to wide interpretation by the courts. Yet, as weak as it was, Daugherty's bill never made it out of the House Conservation Committee. Two state senators from eastern Ohio proposed similar legislation in 1939 and 1941, with the added requirement of a performance bond equal to or greater than the tax assessment valuation of the land to be mined, but the bills also died in committee. Following these initial regulatory efforts, war-created demand led to the spread of unregulated coal surface mining and a dramatic increase in production.[10]

With peace declared in 1945, there was a renewed attempt to impose controls on the industry. In that year the state legislature considered five different bills to regulate strip mining, each of which were introduced by legislators from eastern Ohio counties. Only one of the proposals reached the House floor, where it was debated and amended. This measure was sponsored by Representatives J. A. Gordon of Cadiz (Harrison County), Gilbert N. Frash of Malta (Morgan County), C. A. Craig of Cambridge (Guernsey County)— all Republicans—and C. T. McCort, a Democrat of St. Clairsville (Belmont County). Their bill required operators to post a bond of $100 per acre, level mine sites to "pre-existing contour," and make a quarterly deposit of six cents per ton of coal mined with the county auditor, a small part of which was to fund reseeding. In June, when the proposal came up for debate on the House floor, amendments halved the bond requirements, weakened replanting provisions, and established safeguards for operators to the right of appeal from administrative decisions. Yet even in this watered-down form the bill was voted down fifty-eight to fifty-three.[11]

Ten days after defeat of the Gordon proposal the legislature passed Senate Bill 344, establishing a Strip Mining Study Commission (SMSC) to investigate the need for controls on the industry and to draft another regulatory bill based on their findings. Some of the members appointed to the commission had ties to strip mine operators, while others were proponents of strong regulation, including *Cadiz Republican* editor Milton Ronsheim and former

Ohio Farm Bureau legislative director Edwin J. Bath. The members called their first meeting in October 1945, and conducted a number of inspection tours the following month in fifteen Ohio counties as well as areas in West Virginia and Pennsylvania, states which had already passed control legislation. Formal hearings before the commission began in December, in Columbus, providing an opportunity for both opponents and proponents of regulation, as well as supposedly impartial experts in the fields of agronomy, conservation, forestry, and taxation, to make their case. Twenty-five witnesses spoke in favor of regulation, sixteen appeared in opposition, and fourteen gave "neutral" testimony. Although the tours and hearings did little to change the views of individual commission members, in the following months they agreed on the draft of a report that included another control measure.[12]

The SMSC bill regulated surface mining operations producing 250 tons of coal or other minerals in one year. Mine or quarry operators were to secure a license for a small fee, put up a bond of $100 per acre with a $1,000 minimum and a five-year liability, cover all exposed coal seams with a minimum of 3 feet of earth, bury all pyritic shale, seal off any breakthrough to underground workings in a coal seam, provide access roads for fire control, level off peaks and ridges of spoil banks to a minimum width of 15 feet cross section, and plant a cover crop of trees or grasses on the banks. Additionally, the commission recommended that the chief of the Division of Mines be charged with licensing and regulating the strippers, and that the director of the Agricultural Experiment Station be given authority to administer all planting and any other reclamation. It also suggested passage of a general severance tax on all mineral production, to finance a general program of conservation. The recommendation on leveling caused some controversy among commission members, but otherwise they reached a consensus on the provisions of the proposed legislation.[13]

The SMSC delivered its report to Governor Herbert and, in January of 1947, he presented it to the General Assembly and urged passage of the recommended control bill. Strip mining could not be prohibited, he said in his address to the legislators, but "it would be the height of folly for the people of Ohio to spend large sums of money to conserve natural resources and at the same time to permit unscrupulous operators to mine coal in a manner which is in complete conflict with the general policy of conservation." Bills were then introduced in each chamber, including one by Clingan Jackson, a senator and member of the commission, and another by Ray White, a member of the House and Democratic newspaper editor from Holmes County. Their proposed legislation was virtually identical except for the size of the bond and the financial liability of the operator in terms of revegetation. Jackson's bill called for a bond of $100 per acre and a replanting liability of $50 per acre, while White's bill set a $200 bond and $100 replanting obligation. In the House, efforts to lower the size of the bond failed, but amendments exempted operators mining less than 250 tons per year and allowed for replanting to be waived by the director of the Agricultural Experiment Station. When debate moved to the Senate Conservation Committee, Evert E. Addison, a Columbus (Franklin County) Republican and chairman of the Strip Mining Study Commission, offered two amendments to eliminate the provisions that required coal operators to level spoil banks and cover exposed coal seams. Committee members rejected these changes but they did lower the bond to $100 per acre and set replanting liability at $50 per acre, in line with Jackson's bill and the recommendations of the commission. Addison and several other lawmakers then walked out of the meeting before the committee members recommended passage of the modified act, House Bill 314, on a six to zero vote. In June, the full Senate passed the proposed legislation on a thirty-two to two vote, with one of the "nays" cast by Carl Schurtz of Coshocton (Coshocton County), who took the position that the bill was too weak. The House concurred with the changes made on June 14, and the conference bill went to Governor Herbert for his signature.[14]

In addition to setting bond amounts, defining the liability for revegetation, and exempting small operators, House Bill 314 included provisions explaining the need for the law and elaborating on what would be required of operators. The preface to the regulatory components deemed the act

> an exercise of the police powers of the state for the public safety, the public health and the general welfare of the people of the state by providing for the conservation and improvement of areas of land subjected to strip mining; by aiding thereby in the protection of wildlife; by decreasing soil erosion and flood hazards; by aiding in the prevention of the pollution of lakes, rivers and streams; by decreasing fire hazards and by preventing combustion of unmined coal; and by generally improving the use and enjoyment of such lands.[15]

The law was to become effective January 1, 1948, after which all coal stripping companies would be required to apply to the chief of the Division of Mines for an annual permit and pay a $50 fee. The penalty for operating without a permit was $100 to $1,000, with each day counted as a separate violation. Operators were also required to submit a detailed report one month after stripping began and six months after completion of the mining. Bonds would be released after satisfactory restoration, as determined by the Experiment Station director on inspection of the site one year after leveling and replanting. Restoration standards were similar to the practices specified in the SMSC report, including leveling off peaks and ridges of spoil banks to a minimum width of 15 feet cross section; planting trees, shrubs, or grasses (as recommended by the director of the Experiment Station); covering coal faces; and sealing breakthroughs to deep mines. Forfeited bonds, including those covering mines where restoration was not completed within five years after mining ceased, were to be deposited in a reclamation fund. Monies of the fund would be made available to the Agricultural Experiment Station for restoration of abandoned sites.[16]

"The People" and Their Arguments

During the process leading up to passage of House Bill 314, proponents of regulatory legislation promoted their position in various ways, both as individuals and as groups. Milton Ronsheim, the editor of the Harrison County *Cadiz Republican*, was a spark plug for the control campaign. Throughout the 1930s and 1940s he routinely published stories and editorials on surface coal mining in his paper, he convinced Frank Lausche to take up the issue of strip mine regulation during Lausche's successful bid for governor in 1944, and he served on the Strip Mine Study Commission. Ronsheim tended to see the effort to pass a strip mine bill as a contest between vested coal interests and common people typically locked out of the political process. "The demand for legislation to regulate coal stripping has come from the people," he declared, "with no money, no organization and no lobbyists in Columbus, such as the operators have." The campaign was "a cry for self preservation."[17]

Once aware of the degradation caused by surface coal mining, Lausche himself began a personal crusade for regulation. Speaking in Cadiz, Ohio, in the summer of 1945, he described strip mining as "sheer butchery, disemboweling of the land and leaving its ugly entrails exhibited to the naked eye." The hillsides and streams, Lausche declared, "must be kept as the homes of the people." During his first term as governor, he played an instrumental role in passage of the bill establishing the study commission and he selected its public members. His appointments to the SMSC—including Ronsheim and Bath—were ardent supporters of regulation of the strip mining industry and their presence certainly made for stronger legislation in the end. Lausche also set the tone for the work of the commission when, prior to the organizational meeting in October, the members met in his office and he told them that no problem in Ohio was "more acute and more deserving of attention" than the damage from coal surface mining.[18]

The most important support in the campaign for regulatory legislation, however, came from farmers' organizations, particularly the Ohio State Grange, Ohio Farm Bureau, and their respective county chapters, as well as rural groups supportive of farmers' concerns. Most of those who offered testimony before the study commission in December 1945, and at Senate hearings in March 1947, were farmers' representatives. Leaders of other organizations, including local sportsmen's groups, a statewide association of rural pastors, and the state organization of county commissioners, echoed the sentiments and arguments of the farmers while also raising issues of their own. In addition, members of farm families and individuals from farming communities participated in lobbying for passage of strip mine legislation by writing letters and sending telegrams to the governor and state legislators.

The arguments the Ohio farmers and other rural-dwellers made for regulatory legislation in the 1940s can be grouped into three categories: economic, spiritual, and aesthetic. First and foremost, proponents of controls on stripping were interested in protecting the agricultural productivity of the land, not only to sustain individual material gain but also to preserve the integrity of rural life. As one Whipple, Ohio, resident simply put it, "I think our country would be better fit for farming." Those in favor of regulation maintained that farmers worked the soil with a long-term perspective and their ability to continue to use the land productively was intimately linked to the stability of local communities. Strippers, on the other hand, were purportedly motivated by greed and had little interest in either the health of the soil or the well-being of the surrounding communities. In addition to this argument, support for state regulation was also based on a belief in the land as a creation of God, which carried with it a mandate for stewardship. Control advocates portrayed strippers as violators of this mandate, desecrators of sanctified land, while they counterposed farmers as the more legitimate caretakers of the natural world. And finally, the push for regulatory legislation was a matter of aesthetics. The gouging of the earth to get at coal seams violated the beauty of the land. Unencumbered by any law requiring reclamation, the farmers and other rural dwellers maintained, mine sites were left as barren testaments to strippers' disregard for the aesthetic pleasures of the hills.[19]

The most common argument proponents of regulation made concerned agricultural productivity and competing land use, as well as the links these had to community stability. In 1943, the Ohio State Grange Conservation

Committee noted that strip mining left "the land in a condition which is untillable and not suited to good agriculture and conservation practices," and it called for cooperation with unspecified agencies in solving this problem. Three years later the committee expressed opposition to "the needless destruction and wanton waste of natural resources being brought about by strip mine operations," and it recommended immediate remedial legislation. In 1947, the deputy master of the Morgan County Grange, representing two hundred members of twenty-two local Granges, explained his county organization's position. "We believe that strip mining is a menace to the agriculture and the very life of our county," he wrote, "unless some control measure is taken." Like all other eastern Ohio County Grange and Farm Bureau chapters, as well as the state organizations, the Morgan County Grange favored House Bill 314 to save eastern Ohio agriculture. But even sportsmen's organizations emphasized the harm strip mining did to good farmland. "The membership of the Western Tuscarawas Game Association," its president explained to Governor Herbert, "feel that in the interest of the State of Ohio, our County and the heritage to be handed to our children, there must be some regulations provided for the Strip Mining of Coal . . . Strip Mines must level their Spoil Banks and the land put in a tillable condition."[20]

Linked to the argument about land use was the notion that farming, unlike coal strip mining, sustained communities. A petition transmitted by John E. Thompson, pastor of a church in Beverly, Ohio, made this connection. "We the undersigned," it read, "view with deep concern the vast strip mining operations being carried on in Noble, Guernsey and Harrison Counties. These operations [are] resulting in the disfigurement of the country side—causing valuable acres of farm lands to be taken out of agricultural productivity—reducing the population of these communities, and creating serious economic conditions in these communities." Farmers' organizations and rural eastern Ohioans suggested that farming was better than strip mining for the community because it involved a long-term interest in the land, as opposed to a desire for short-term profits that was supposedly characteristic of strip operations. A hardware store owner in Athens County explained, "If the income from a ruined farm could be computed over the next hundred years in dollars and cents it would be many more times than the few *quick* dollars made by destroying absolutely the land." Other critics dismissed the claims of surface miners that they were motivated by more than simple greed. Attempting to dispel the notion that strip operators had increased production during the war as an act of patriotism, a Perry County commissioner and small-scale stripper told the Strip Mining Study Commission that "they did it for profits and nothing else." Somewhat along these lines, *Morgan County Herald* editor W. D. Matson brought the developing Cold War with the Soviets into the debate. "People of Morgan county," he wrote, "the Russians are not going to buzz-bomb us—we are to be mined out of our homes and our Fatherland despoiled by American coal barons who seem to care nothing for aught but the dollar sign."[21]

Proponents of regulation also made an argument that God had given people the land and its resources to use but not to destroy. Many called strip mining a crime against God, including the hardware store owner quoted above. These same control advocates also suggested that failing to be Christian stewards by disrupting the harmonies of the God-made land would bring retribution. As would be expected, rural pastors from eastern Ohio were most forceful in making such spiritual arguments against strip mining. One resolution sent to the state Public Affairs Department by pastors from Harrison and neighboring counties stated,

> Believing the Creation of the Human Family and the Sacred Soil to be the Divine work of our Father God; Believing that Life and the Fruits of the Soil are inseparable and that each are a trust in our hands from God; And recognizing the extensive and unchecked destruction of great areas of farm pasture land by the strip mining of coal, with its consequent undermining of community life and land values for generations to come in those areas of our State; We again advocate the control and regulation of strip mining by our state government.

From the pastors' point of view, farming was not a threat to a natural, God-designed equilibrium, but strip mining certainly was. Some of the same religious leaders also used their moral weight and positions of influence to make nontheological arguments against surface coal mining. In January 1945, with 2,300 Protestant Ohio pastors registered, the Annual Ohio Pastors Convention unanimously passed a resolution demanding regulatory legislation. Controls were needed, the resolution explained, "to provide for public health and safety and to properly protect both private and public property rights and to insure the usefulness of land" impaired by stripping for coal and other resources.[22]

Farmers and other rural people in eastern Ohio were also proponents of regulation because they viewed strip mining as destructive to aesthetic qualities of the landscape. Speaking at a Harrison County Farm Bureau picnic in 1945, representing the state organization, Herbert Evans pointed to the surrounding soon-to-be stripped hills and exclaimed, "The farmer says it's all right to take the coal out of these hills but they must not be left an unsightly dump." In a letter to the *Cadiz Republican*, a Harrisville resident described the changed aesthetics she saw on her car trips through eastern Ohio strip mine country. "Until now the ride from Harrisville to Cadiz and over other nearby roads was a pleasure," she wrote. The area had "lovely farms, with the green fields, nice homes and outbuildings, stock grazing, and every thing to make the trip pleasant. Now we go when we have to for it's no pleasure to see where those lovely fields have been turned upside down, farm homes destroyed and in their places those awful unsightly piles of dirt, that never can be restored to their original beauty." For Mrs. Householder and others, surface coal mining meant the disappearance of a beloved and beautiful pastoral scene.[23]

Sometimes strip mine control advocates made all or a combination of these arguments—about agricultural productivity, Christian stewardship, and aesthetic ruin—at once. In 1945, Joseph Fichter, master of the Ohio State Grange, addressed delegates at the organization's annual meeting. "How long will it be before we learn that exploitation of people and of our God-given natural resources does not pay?" Fichter asked. "We have a conservation problem in Ohio which is the result of the exploitation of land and water resources. The defacing of land and the displacement of farm homes through strip mining of coal cause us to resort to demand for government regulation in order to prevent such waste."[24] Though reluctant to bring the state into the problem, Fichter and other conservative farmers saw no other way to deal with the sinful exploitation of the land, destruction of rural communities, and impairment of scenic beauty by surface coal miners. Consequently, they supported passage of the first regulatory legislation addressing the problems of stripping.

But not all the people of eastern Ohio were in favor of controls. House Bill 314 had its opponents, including operators who would have no regulation of the industry, and they were nearly as vocal as the measure's proponents. A few operators, such as Joseph Fay, who had extensive operations in Harrison and Jefferson Counties, appeared voluntarily before the SMSC in favor of controls. Those who supplied strippers with machinery, loans, or a restaurant meal, however, as well as strip mine employees, private truckers who hauled the coal, and most coal operators all protested against regulation. They worried about a loss of business and a loss of jobs if the 1947 bill passed. Responding directly to unregulated strip mining's critics, R. A. Christian, of the Canton Supply Company, explained, "We believe strip mining has proved of more value to the state in the mining of coal and maintaining Ohio industry then [*sic*] could be produced from this same land in 500 years of farming." Private coal hauler Dale Brannon, from Tuscarawas County, pleaded with the governor "to protect my job from those outside the coal industry promoting House Bill 314." Coal operators not too subtly hinted they would close their strip pits down if the regulatory legislation passed, claiming it would force them to operate at a loss. Strip miners themselves, whether voluntarily or under pressure from pit owners, also spoke out against the control bill. R. S. Patterson, the owner of Beaver Fork Coal Company, in Columbiana County, forwarded a petition to Governor Herbert signed by five workers and written on company stationery. "[H.B. 314] is being promoted by persons not interested in the coal industry," the petition read, "and the passage of such a bill can conceivably deprive us of our means of livelihood."[25]

Shifting Forces of Opposition

Once it was signed by the governor, coal operators denounced the 1947 control law and, after it went into effect in 1948, they filed a suit in court to test its constitutionality. Yet within eight months, the Department of Mines reported, all surface mine operators in the state had registered under the law and posted performance bonds. The operators also showed themselves willing to try another approach to getting more lenient control legislation. Frank Lausche was elected governor again, in 1948, and opponents and proponents of the law met at his urging to craft a compromise. One morning James Hyslop, then vice president of Hanna Coal Company, showed up at the *Cadiz Republican* offices to tell Milton Ronsheim that his company was through fighting and "the coal people" wanted an "agreed" bill. Ronsheim cleared some clutter off his desk and the two sat down for the first of several long sessions to draft new legislation. The newspaper editor later claimed the bill they wrote "left much to be desired," but the operators dropped their law suit and agreed to the general principle of reclamation in return for amending the earlier law. Sponsored by Ed Schorr, the compromise measure met no opposition from the coal lobby and, with only minor noncontroversial changes in phrasing in committee, it sailed through the legislature in June and July of 1949.[26]

The new law centralized administrative responsibility, increased the performance bond, and substituted an elastic surface restoration requirement for the leveling provision of the 1947 act. Administration was transferred to a Division of Reclamation, in the Department of Agriculture, with a chief appointed by the governor and confirmed by the state Senate. A five-person

Reclamation Board, also appointed by the governor with the consent of the Senate, was established to hear appeals from any decision of the new division. The board would include one representative of the industry and another who spoke for "the public," as well as three experts in various fields, with all members serving staggered five-year terms. Permits were replaced by a license, which included a fee of $50 and $10 for each acre the operator planned to strip. Bonds, which were to be released progressively, were raised to $190 per acre, with a $1,000 minimum. Operators were given two years to reclaim the land affected by their surface mining, though they had only to return the surface to "a gently rolling topography." They were also freed of the provisions requiring sealing of all exposed deep mine works and covering exposed coal faces. Replanting was limited to those areas "where revegetation is possible," and the act gave the chief discretion to extend the time limits for compliance or waive restoration requirements altogether. The provision allowing for substitute acreage in the 1947 law was retained.[27]

Though Milton Ronsheim played an important role in writing the new legislation, Governor Lausche's attempt to appoint him as the "public member" on the new Reclamation Board was blocked by members of the legislature and strip industry. Ronsheim received an absolute majority in the Senate but he lacked the votes for the necessary two-thirds confirmation. The governor also tried but failed to appoint C. C. Fay, one of the state's few pro-regulation strip miners, as the industry representative to the board. Fay was rejected by the Ohio Coal Association, for his reputation as a "reclamation operator," and by the United Mine Workers, for running a nonunion surface mine. His appointment was turned down by a majority of senators. But Lausche did manage to appoint Wilbur Matson, the editor of the *Morgan County Herald* and associate of Ronsheim. His choice for chief of the Division of Reclamation, Zoyd M. Flaler, was also acceptable to enough senators for successful confirmation. Flaler was a former highway engineer and new to the debate over regulation. Once on the job, he took a conciliatory approach to enforcement. The division sent out thirty-five cease orders for noncompliance with the new law by the early part of 1950, but these were generally restricted to small operators and in most cases violations were quickly corrected when identified.[28]

In the decades that followed, several amendments were made to the 1949 law, often in response to pressure from the Ohio Grange and the Farm Bureau. By the early 1950s, the Conservation Committee of the Ohio Grange was bemoaning the lack of enforcement of the regulatory legislation, and its rhetoric echoed earlier complaints. Strip operators had ruined the productivity of 2 million acres of lands formerly devoted to farming and forestry, the organization claimed, and since most of these lands were devalued for tax purposes, they had become financial liabilities for those counties where they were located. As a solution to these problems, the Grange advocated transferring administration of the control law to the newly created Division of Lands and Soils within the Ohio Department of Natural Resources (ODNR). In 1953, the Conservation Committee claimed that "land in the strip mining areas of Ohio are [sic] left in such condition that it is practically worthless," and called for leveling of land by the operators for farming where practicable and reforestation where it was not. The division was transferred to the ODNR, but at the end of the decade the Grange was still demanding that the agency "require the strip mine operators to improve the restoration and leveling of their spoil banks so as to leave them in good condition for seeding and planting trees."[29]

The Ohio Farm Bureau also continued to advocate better enforcement and improved control legislation. In 1965, the legislative representatives for both the Grange and the Bureau lobbied the Ohio General Assembly, leading to an overhaul of strip mine regulations. But the new control law included an only slightly less vague requirement to grade spoil banks "so as to reduce the peaks thereof and reduce the depressions between the peaks of such spoil banks to a gently rolling, sloping, or terraced topography, as may be appropriate, which grading shall be done in such a way as will minimize erosion due to rainfall, [and] break up long uninterrupted slopes." The act also charged operators with clearing the surface of large rocks or other obstructions, "as may be appropriate," to permit the operation of machinery and make the area more suitable for revegetation. In addition, it included a provision requiring the prevention of acid mine drainage and siltation of streams, "if possible," on adjoining lands. Despite this latter amendment, in 1966, delegates to the annual Ohio Farm Bureau meeting felt compelled to adopt a policy statement declaring their intention "to initiate legislation for the protection of landowners adjacent to an area licensed for strip mining who may be affected by excessive siltation from the mining operation." The delegates had received reports of damages resulting from silt and clay washing onto lands adjoining strip mines to depths of up to 2 feet, and of the difficulties farmers had in getting reimbursement. Following passage of the resolution, the Farm Bureau sponsored a bill granting the chief of the Division of Reclamation the right to deny a license to strip mine operators when he judged the operation may result in damage to neighboring land. But two years later, the farmers' organization was still insisting on strict enforcement of reclamation laws and, like the Grange, calling for the broadening of controls to cover other types of surface mining besides coal.[30]

Actually, by the late 1960s, Ohio farmers and their designated representatives were being displaced as the primary advocates of surface mine controls. As agriculture declined in the eastern Ohio strip mine counties many of the new opposition leaders in the state were self-styled conservationists and environmentalists. Their concerns were somewhat different from those of the farmers and they were increasingly interested in taking the fight to the federal level. At a 1970 strip mining symposium in Cadiz, Ohio, most of the four hundred participants were students, and symposium sponsors included the Ohio Conservation Foundation, the League of Ohio Sportsmen, the Ohio Audubon Council, and the Sierra Club's state chapter. When the Hanna Coal Company expanded its stripping operations in Belmont County in 1972, it drew opposition from the newly formed Citizens Organized to Defend the Environment (CODE) and the somewhat older Appalachian Ohio Research and Information Group. The two groups articulated a critique of Hanna's strip mining that blended familiar economic concerns with increasingly prevalent environmental concerns. They objected to the subordination of the public good to private interests, but they were also alarmed "at the serious disruption of the intricate ecological balances, the stability of which were produced only through millions of years of natural processes."[31]

Members of CODE and other Ohio groups concerned about surface coal mining also became disillusioned with state remedies, a process experienced by others at about the same time throughout Appalachia. Beginning in the early 1970s the various state activists began to create a regional movement for federal action, with some proposing regulatory legislation and others—such as Ohio's Concerned Citizens Against Strip Mining (CCASM)—demanding abolition. CCASM was formed in 1970 by John Deboins, vice president of the Belmont-Marion AFL-CIO, and Claude Colvin, a linguist at the Ohio State University Marion campus. In the summer of 1971, the new group sponsored a meeting that drew prominent strip mining activists from nearby states as well as the eastern representative of the Sierra Club. That October, opponents from Kentucky, West Virginia, Ohio, and Virginia also met in Huntington, West Virginia, to form the Appalachian Coalition, which would coordinate efforts to ban stripping at the national level. And, in June 1972, nine hundred activists gathered at the Environmental Center of Union College, in Middlesboro, Kentucky, for a National Conference on Strip Mining. By the early 1970s, there were few opponents of strip mining seriously considering state-level action to control or ban strip mining. The attention of most activists was focused on the many regulatory and abolition bills then before Congress. But the early concern expressed by the likes of Ohio farmers had been an important step forward in the campaign for government intervention.[32]

Selfish Interests

Opposition to Surface Coal Mining in Pennsylvania

Although farmers were eventually displaced as the leading opponents of unregulated surface mining in Ohio, the history of their early efforts suggests that people working on the land could be advocates of the complex mix of conservationist and preservationist ideas that have characterized U.S. environmentalism. Similarly, the history of a 1961 campaign to regulate the Pennsylvania strip mining industry points up the contribution of industrial workers to the development of modern environmental activism. Yet that campaign and efforts to establish stricter controls in the years that followed also reveal some of the complexities of "labor environmentalism." Initially, Pennsylvania workers and sportsmen cooperated with one another and rejected operators' attempts to counterpose jobs and the environment. They seemed to have a common objection to the destruction of the land for exorbitant profits and this was the basis for a nascent labor-conservationist alliance. Sentiment was perhaps not always genuine, however, and the alliance was not the most hardy of coalitions.[1]

This chapter focuses specifically on the role played by the United Mine Workers of America (UMW) in the 1961 campaign. Leaders of that union expressed at least nominal support for strict controls early on in the legislative process. But their commitment to strip mine regulation was largely a matter of expediency, a way of hampering the production of an industry hitherto resistant to organization. When it appeared that no control bill would pass, local officials intervened to save a compromise measure that sportsmen's groups opposed, pledging to work for a stronger control law in the next legislative session. They did not return, however, and it was left to the sportsmen alone to pass a better bill. After the 1961 campaign, in fact, the union was more of a hindrance than a help to the passage of meaningful strip mine legislation, in Pennsylvania and other states as well as at the national level. Through the rest of the decade union leaders as well as some of the rank-and-file membership tended to speak and act contrary to the expressed interests of other common people, including deep miners, when the protection of surface mine jobs appeared to be in conflict with protection of the environment.

Strip Mine Controls in Pennsylvania, 1945–1961

Pennsylvania passed its first legislation controlling bituminous coal strip mining in 1945, "to aid thereby in the protection of birds and wild life, to enhance the value of such land for taxation, to decrease soil erosion, to aid in the prevention of the pollution of rivers and streams, to prevent combustion of unmined coal, and generally to improve the use and enjoyment of said lands." Like other state regulatory acts, the Pennsylvania law required operators to put up a bond ($200 per acre) and stipulated the reclamation standards to be met for the bond to be released. Within one year after an operation was completed, the strip operator was required to cover the face of unmined, exposed coal and to level and round off spoil banks "to such an extent as will permit the planting of trees, grasses or shrubs." Replanting was to occur within the same period of time, although an extension could be granted by the secretary of forests and waters, and replanting on other previously stripped areas could be substituted for this requirement. Any forfeited bonds would be deposited in a Coal Open Pit Mining Reclamation Fund, and operators caught mining without a permit would be liable to pay a fine not exceeding $500 (offering little in the way of deterrence).[2]

Soon after passage of the 1945 law, coal operators challenged the act in court on the grounds that it discriminated against them by singling out their industry. Coal strippers had won a similar lawsuit in Illinois, overturning the 1943 control law there. When *Dufour v. Maize* found its way to the Pennsylvania Supreme Court, however, the justices ruled against the strip mine operators, holding that the Bituminous Coal Open Pit Mining Conservation Act neither violated the constitution nor deprived coal companies of property without due process of law. The classification of "open pit mining" as

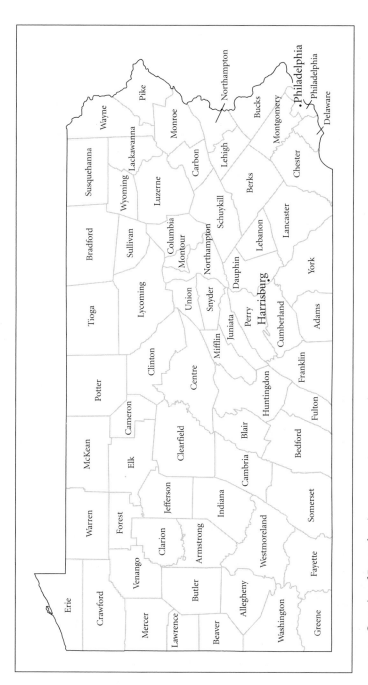

MAP 3. Counties of Pennsylvania

distinct from other mining was founded on real distinctions, the justices declared, and the statute should stand. Yet the courts were not the only means coal operators had at their disposal to slow or block the implementation of regulations on the industry. As in other states the initial law was weak and, without a specially designated government agency administering the act, enforcement was lax. As a result, despite operators' failed legal challenge, surface coal mining in Pennsylvania went virtually unregulated, devastating thousands of acres of land and contributing substantially to the worst acid mine drainage problem in the country.[3]

The failure of the 1945 act to adequately regulate the strip mining industry eventually prompted various groups to organize for another control bill. Leading the effort for a new law was the Pittsburgh-based Allegheny County Sportsmen's League (ACSL), a branch of the Pennsylvania Federation of Sportsmen's Clubs (PFSC). Sportsmen were witnesses to the ruination of trout streams and habitat for wild game by surface coal mining and they sought better regulation to preserve the sites necessary for their recreational activities. The ACSL was assisted in this by the Allegheny County Labor Committee, which included the United Steel Workers, the International Brotherhood of Electrical Workers, and the United Mine Workers. There probably was, in fact, some crossover between the organized sportsmen and organized labor, with steelworkers, electrical workers, and miners joining sportsmen's clubs as they took up hunting and angling in greater numbers during the postwar period. These working-class sportsmen were particularly concerned about the profits strip operators made in the process of destroying the Pennsylvania landscape. Their class position and recreational interests allowed them to see the connections between strip operators' greed and their lack of concern for the environment.[4]

In 1961 the various opposition forces joined together to introduce House Bill 1438, which provided for an increased bond of $500 per acre, backfilling of strip pits, operator responsibility for acid mine drainage (even when it originated from abandoned deep mines), and the use of forfeited bonds to correct stream pollution in strip mine regions. Near the end of June, after a failed attempt at weakening the bill with amendments, the Pennsylvania House passed it on a vote of 150 to 51, over the vehement protests of legislators from strip mine counties and strip coal industry lobbyists. Representative Paige Varner (R-Clarion) contended that the bill "frustrates Pennsylvania's efforts to hold its industrial payroll by advancing mining costs to the point where jobs are endangered." When Representative Austin J. Murphy (D-Charleroi) suggested that controls would merely reduce operators' profits, forcing them to shift a few men from digging coal to filling holes, Varner replied by setting the health of the industry against the quality of the environment. "I don't see why we should give up a growing industry," he said, "for a few fish or a few trees." This was rhetoric meant to drive a wedge between conservationists and working people, forcing them to make a choice between nature and their livelihoods. The lobbyist for the Central Pennsylvania Open Pit Mining Association (CPOPMA), W. G. Jones, tried a variation of that approach. "[I]t would be unfortunate," he threatened, "if the one and a half million acres of hunting land owned, leased or otherwise controlled by the open pit industry were to be posted against hunting this coming fall."[5]

Although they lost the vote in the House, opponents of new regulatory legislation scored a victory shortly after when H.B. 1438 and an identical companion bill covering anthracite coal were referred to the Senate's Local Government Committee, chaired by John J. Haluska (D-Cambria). Normally, the bills would have gone to the Mines and Mineral Industries Committee or the State Government Committee. The only reason to send it to Local Government, according to Senate observers, was to kill it. Senator Haluska represented a county in central Pennsylvania with many active surface mines and he was a close personal friend of C. E. Powell, one of the state's largest strip mine operators. Haluska handled Powell's insurance policies and, soon after his committee received the strip mine bills, he was shown on the front page of the *Pittsburgh Press* driving a car licensed in Powell's name. When H.B. 1438 and its anthracite companion came to the Local Government Committee, Haluska announced that Daniel C. Parish, an influential Allegheny contractor and close friend of Governor Lawrence, had asked him "to go slow on this because it would hurt industry and make people lose their jobs." Doing Parish's bidding, the senator refused to release the bills for most of summer.[6]

In the face of this intransigence, supporters of strong control legislation continued to speak out and mobilize. At the end of June the executive board of Westinghouse Local 601, a United Electrical Workers affiliate, unanimously passed a resolution calling on the Senate to follow the House in quickly passing the two control bills. "The work of a greedy, small group of strip miners is designed to make Pennsylvania a desert wasteland," it read, and "obstruction to good legislation to control strip mining must be stopped." When Haluska announced hearings on the bills to begin in July, the PFSC began coordinating a lobbying effort by its members as well as a motor caravan of sportsmen, conservation groups, women's clubs, and garden clubs to Harrisburg, the state capitol. John Laudadio, the legislative chairman of the federation, said he knew the hearings were a stall tactic but his organization was encouraging attendance and participation anyway. Just prior to the hearings, state mining secretary Lewis E. Evans set the tone for the debate to come by calling the existing strip mine controls "virtually useless." Under the law, he said, land had to be recontoured only when mining was "completed," a term indefinite enough to allow operators to delay backfilling for as long as ten years.[7]

On the first day of the hearings, July 10, Senator Haluska immediately created an uproar in the chambers by implying a $25,000 payoff had been made to legislators for passage of a strong regulatory bill. Proponents of a new law ignored the charge and proceeded to explain their position. Those testifying included the secretary of the state Department of Mines, a representative from the League of Women Voters, the legislative chairman of the PFSC, a pro-regulation stripper, the legislative chairman of the ACSL, and the master of the state Grange. Representing the ACSL, Frank Shean denounced strip mine operators as being more interested in "making a fast profit than safeguarding the public," pointing out that four children had died in western Pennsylvania strip mines since mid-May. "With every broken body taken from a strip mine," he said, "somebody made a few extra dollars." J. Collins McSparran, master of the Pennsylvania State Grange, echoed the sentiments expressed by Ohio farm leaders in the 1940s and 1950s. Pennsylvania farmers were not opposed to strip mining, he explained, but they were vigorously opposed to the manner in which many operations had been conducted, particularly the complete disregard of the future use of the land. In many communities, operators had destroyed the productivity of forest and field and simply moved away, leaving spoil banks, polluted streams, dangerous pits, and depressed property values in their wake. Careless and incompetent operators should not be permitted "to convert the scenic beauties of our countryside into ugly barren wasteland, and to create rural slums," McSparran argued, "just because [they] are unwilling to fulfill their moral responsibilities." Later that night, these arguments were driven home with some force when a power shovel and bulldozer valued at $175,000 were destroyed by dynamite at a Fayette County strip mine. No one was injured, but the blast touched off a fire visible for several miles.[8]

The next day's witnesses at the hearings included opponents of stronger regulation. Joseph Baran described himself as a self-employed strip miner and member of UMW Local 6326. He charged that the control measures were part of "a conspiracy between the big coal companies, the mines department and the United Mine Workers to put the nonunion operators out of business." W. J. McCabe claimed to represent six Cambria County sportsmen's organizations and questioned the legitimacy of the PFSC leadership on the strip mine issue. He repeated the earlier threat made by the CPOPMA lobbyist that hundreds of thousands of acres of hunting areas would be closed to sportsmen if the control bills were passed and, in effect, challenged the notion that hunters and anglers and workers had common interests. But McCabe made an appeal for the unrestricted extraction of natural resources too. Part-owner of a lumber firm, his property also had coal under it and, he worried, "if they can stop me from mining it they can stop me from cutting trees, just because they think they're beautiful."[9]

McCabe's clumsy attempt to drive a wedge between working people and conservationists was not very successful. The day following his testimony, outside of the hearings, the Allegheny County Labor Committee announced its full support for strip mine legislation. The Labor Committee served workers as an independent political voice, unassociated with either the Democratic or Republican parties, and its leaders recognized the connections between workers' interests and conservation. "We are unequivocally opposed," the organization declared, "to the selfish interests and to the legislators they appear to control who resist effective regulation of strip mining, [which] despoils our natural resources and endangers the health and lives of our citizens." The committee also notified sportsmen's groups that they stood with them "in conservation of Pennsylvania's forests, fields, lands and running waters."[10]

Perhaps in response to the Labor Committee's resolution and its expression of solidarity with conservationists, coal operators and their allies made even greater efforts to counterpose the health of the economy and environmental quality during the rest of the hearings. In his testimony, a representative of the Pennsylvania Railroad put the company on record as opposing the strip mine control bills on the grounds they would raise operating costs for strippers and affect the coal hauling business. The corporation's manager of coal traffic explained that if strip mining was cut back 80 percent, as strippers estimated the effect of the regulatory bills, the railroad "would have to lay off hundreds and perhaps thousands of employees and could lose as much as $12,800,000 in revenue." Speaking for the CPOPMA, attorney Joseph J. Lee claimed tougher backfill requirements would put more than 1,100 miners out of work in central Pennsylvania alone. Trying a more folksy approach, the president of the Sunbeam Coal Corporation described sportsmen as utopians with a dim view of reality. "Food on the table," he said, "is more important than the delights of fishing."[11]

After the last of seven nonconsecutive days of testimony, Senator Haluska declared that his committee would not take any action on the control bills until the transcript of the hearings was prepared. Once that was done, the "strip mine problem" would be turned over to a fact-finding committee. This was more stalling, an attempt to drag out consideration of the bills to the end of the legislative session, and the next day Haluska received a death threat which read:

> I am a good sportsman and I love the outdoors. I hate cheats and crooked politicians. Unless you act on the strip mining bill within two weeks, by Aug. 8, 1961, you might die rather suddenly. You may think this is a prank, but believe me this is no joke. Sincerely, an honest sportsman who is a good shot with a gun.

Whether a prank or serious threat, plainclothes state police were attached to the senator and, the day he revealed the letter, Haluska also announced that the Local Government Committee would vote on the proposed legislation within two weeks, just meeting the deadline set by the sportsman. In the meantime, committee members would conduct a two-day tour of strip mine counties in both the bituminous and anthracite regions of the state.[12]

On the first day of the tour, August 2, Senator Haluska and eight of the other eighteen members of the Local Government Committee gathered at a supper-club motel outside the town of Clarion, in the heart of bituminous coal country. They were met by Charles Leach, a former coal stripper, Clarion County GOP finance chairman, and insurance agent who specialized in writing bonds for strip coal operators, as well as Milt Breniman, a Clarion real estate agent and the tour leader. Two sportsmen also traveled with the group, and they reported that the itinerary seemed to be representative of Clarion conditions. A few days later, however, John Laudadio and William Guckert sent an open letter to Senator Haluska asking why state mines officials, conservationists, and "the people" were not represented on the tour. Guckert claimed to have telegraphed the senator earlier in the week, asking for an invitation and proposing to take the group through northern Butler County, but he had not received a response. The two sportsmen also pointed out that some of the worst conditions in the state were no further than 15 miles from where the committee was being shown what the strip coal operators wanted members to see.[13]

The second day of the tour was given over to travels through the anthracite coalfields around Pottsville (Schuylkill County), and the highlight of the trip came at the top of a strip mine spoil bank there. When Senator William Lane (D-Washington) suggested that the tour might be rigged he nearly precipitated a fistfight with Haluska, who then proposed an amendment to weaken the bills before his committee. He would accept a 45-degree backfill requirement where the land is productive, "on farm land and the like," in exchange for knocking out the provisions calling for 45-degree backfilling on nonproductive land, such as "here in the mountains where it is impossible." By the end of the day a deal had been struck, and there was general agreement that strict backfill requirements in the anthracite region should apply only to operations near roads or residential areas. Having gotten some of what he wanted, Haluska returned to Harrisburg and announced that the Local Government Committee would soon release amended strip control bills to the full Senate. Under the new plan, the state Department of Mines would be granted the authority to decide whether the land a strip operator wanted to work was productive, which would be backfilled at a 45-degree angle, or nonproductive, which need be backfilled only to a 75-degree angle.[14]

Meanwhile, letters to the editor of the *Pittsburgh Press* indicated the level and character of the developing opposition to surface coal mining. Some of the letters focused on the profits strippers were making at the expense of the aesthetic quality of the landscape and nearby communities. "Why should a few businessmen be allowed to reap a harvest of profit," Michael Dunny asked, "and in so doing have an utter disregard for the natural beauty of the area in which such operations take place[?]" Mrs. H. Glenn Benton, of Mount Washington, recalled her father's refusal to sell coal rights to strip miners despite neighboring property owners agreement to sell. His "green, beautiful, and productive" fields were soon surrounded by "barren dirt and rock and dirty water holes." Mrs. Benton could not understand how strip miners could "change" the beauty of the countryside this way, and she begged the newspaper to "keep up the good work until enough people get incensed about this to pass laws to see that our God-given heritage is preserved for posterity." Other letters specifically addressed the economics of strip mining. The Mount Carmel Taxpayers Association, in Northumberland County, cited the difficulty of drawing industry to strip-mined areas as one of the most significant problems with the mining method. "It is a detriment to the community," the association officers explained, "and makes it almost impossible for civic organizations to induce industry to locate here." And in mid-August 1961, R. Jordan of Clarion wrote to complain about the unemployment caused by surface coal mining. "Before the advent of strip mining in Clarion County," he (or she) wrote, "there were many more persons employed in producing coal by the deep mine method. . . . Today there is not one deep mine of any importance."[15]

Though unreceptive to these or other arguments, the Local Government Committee finally approved the amended strip mine control bills on a fifteen to one vote, and they went to the Senate floor on August 8. In addition to creating two classes of strip mines, as requested by Senator Haluska, the measures raised the performance bond to only $360 per acre, rather than $500. Even Governor Lawrence thought the amended bills were "too weak," and legislators immediately began an effort to restore them. The following week, senators introduced compromise amendments to the bituminous bill after a conference with the governor and his aides. These changes would make the bill stronger than the one reported out by Haluska's committee but weaker than the proposals supported by conservationists. Bonds would be set at $400 per acre, nonproductive land had to be backfilled at a 75-degree angle and, in addition to strip mines on productive lands, pits within 5,000 feet of any public building or built-up area would have to be backfilled at a 45-degree angle. A hectic ten-hour session ensued, during which time some senators also attempted to restore the legislation to the same form in which it left the House. The latter was beaten back on a close vote, however, and the compromise amendments prevailed.[16]

The next day, after another last-ditch effort to strengthen the bills, representatives of all the concerned parties met in Governor Lawrence's office and emerged with improved measures to send back to the Senate and then to a Senate-House conference committee. The new amendments required backfilling at a 45-degree angle on any stripping operation within 750 feet of a built-up area and at a 60-degree angle in wilderness areas. In exchange, Secretary Evans and others supporting the administration's bills agreed to knock out provisions barring strip mining within 100 feet of a public road and 225 feet of a home, which they thought might invalidate the measures anyway. But legislators from bituminous coal districts in Cambria, Fayette, and Washington Counties withheld their support from the compromise proposals, still hoping to restore the controls to their original form, or at least to achieve something closer to the earlier versions. It was at this point that United Mine Workers leadership from District 5 played an important role in the passage of regulatory legislation. On August 22, district president Joseph Yablonski brought a crew of union officials to a meeting with the eight holdout assemblymen and afterward they all met with Governor Lawrence. With the legislators backing him up, the union leader outlined the union's demands and the governor agreed to them.[17]

Yablonski and his assistants intervened in the legislative process primarily to put pressure on nonunion strip operators. The union had been attempting to organize surface mine workers since the 1930s but with little to show for it. In 1939, District 5 officials signed a first contract with the Strip Mine Operators of Western Pennsylvania, and two years later this was expanded to include UMW Districts 3 and 4, covering southwestern Pennsylvania. There were many more failed organizing campaigns, however, in the Pennsylvania bituminous and anthracite regions and throughout the strip coalfields of Appalachia. Jurisdictional disputes with a rival union of operating engineers as well as collusion between police and government officials proved to be formidable obstacles. By the end of the 1960s, half of the coal mined in western and central Pennsylvania, a third of that mined in southeastern Ohio, and two-thirds of the coal mined in eastern Kentucky and eastern Tennessee was nonunion. Much of that coal was extracted using surface mining methods, and, seeking to protect the interests of deep miners who were 95 percent of District 5 membership, Yablonski lent his union's influence with legislators to the passage of strip mine controls.[18]

Yet even with the UMW's intervention, the summer-long battle for regulation of surface coal mining was not finished. The latest compromise by Governor Lawrence would not actually restore the bills to their original form and the PFSC and ACSL would not agree to "half a loaf," as they put it. The measures provided "no protection against stream pollution," argued the legislative chairman of the ACSL, "and very little against land destruction by strip miners." Mobilizing for more effective controls, the sportsmen began sending telegrams to House members advising against concurrence. But the governor, state mining secretary, UMW, and a majority of legislators were lined up in support of the legislation. Neither United Mine Workers International president Patrick Kennedy nor District 5 president Joseph Yablonski were fully satisfied with the bills, but they were willing to compromise. "We wanted a better bill," Yablonski later noted, "but at this stage of the game its obvious we couldn't get one so what could we do?" Holding up H.B. 1438 as a good start toward the regulation of coal surface mining, he called for strict enforcement and promised to continue the battle for controls during the 1963 session. With the mine workers' assent, on August 29 the House agreed to both the bituminous coal measure and the anthracite bill by wide margins, while a number of those voting against did so at the request of the sportsmen's organizations.[19]

Conservationist Cooperation
and UMW Obstruction

In the aftermath of the 1961 legislative session, it was the conservationist sportsmen rather than the mine workers who held legislators responsible for their votes and followed up on achieving a better law. By the count of the PFSC, twenty-seven senators and fifty-one assemblymen deserved to be unseated for voting against reform in key instances, and its members distributed a list of those to be ousted after their annual meeting in April 1962. Indicating the threat posed by the sportsmen, the two major political parties added prominent strip mine planks in their platforms, and both of the candidates for governor pledged their support for new regulatory legislation. Speaking to a rally of Pennsylvania hunters and anglers, Republican gubernatorial candidate William Scranton said that the most essential thing needed in the conservation field was "a strip mining law with real teeth." In another campaign statement he defined this as "a law requiring an effective bond against damage to land and water, and prohibiting stripping that results in water pollution under any circumstances." Largely as a result of the political activism of the sportsmen, Scranton won the governor's office, four of the most active pro-industry legislators were defeated (including Representative Varner), and PFSC secretary John Laudadio won an assembly seat.[20]

With supporters of strong strip mine controls politicized and organized, the PFSC had announced in September that the original 1961 House measures were no longer acceptable since only complete backfilling to original contour would properly restore stripped areas. In line with the sportsmen's demands, a bipartisan commission appointed by Governor Scranton in January 1963 produced Senate Bill 176, which required backfilling to approximate original contour, acid mine drainage cleanup, increased bonds, and centralized enforcement. In response to this bill, surface coal miners formed the Pennsylvania Conservation Association and waged a campaign to improve the image of the industry. The new organization barraged the media with images of reclamation successes and scheduled tours of reclamation projects for members of the General Assembly and the governor. Following one of these tours, Scranton publicly hedged on his campaign promises and, a week after that, he had Republican lawmakers introduce House Bill 434, which was comparably weaker than S. 176. In May, sportsmen lent their support to the new bill after legislators reinserted the provision for backfilling to approximate original contour and, in early June, the amended measure passed the House on a vote of 196 to 3.[21]

After passage in the House, H.B. 434 was sent to the Senate Mines and Minerals Committee, where it was threatened once again by weakening amendments, including one removing the secretaries of health and agriculture and the executive director of the fish commission from an eight-person reclamation board. On July 1 representatives of the strip mine industry and William Guckert held a stormy two-hour meeting in the office of Senator Thomas Ehrgood, chairman of the Mines and Minerals Committee, and the next day a modified bill was reported out to the full Senate. Guckert claimed it was now even stronger than S. 176. "It will, if properly enforced," he said, "end much of the devastation of our land and the pollution of our streams." In addition to the contour backfilling provision, with variances allowed in some mountainous areas, the measure required strip operators to have a license, set a fine of $500 to $5,000 and up to six months in jail for violations of the law, and established a five-person Land Reclamation Board to oversee operations in the state. The Senate easily passed this bill on a forty-eight to two vote, with one of the "nays" coming from a strip mine county Republican who believed the legislation was unnecessary and the other coming from a Philadelphia Democrat who thought it was not strong enough. In fact, many of the Democrats who voted for the bill also believed it was weak. One called it the flag of surrender by Governor Scranton and Pennsylvania conservationists. On the other hand, Senator Haluska voted for the measure despite his conviction that it "will create chaos . . . and put 100,000 men out of work and on relief before we're through."[22]

In mid-July the House concurred with the Senate version of the bill and the governor signed it into law. Highlighting what had been a central issue in the 1961 and 1963 legislative battles, Scranton explained that his administration had taken action "to sponsor a measure to stop devastation of our natural resources, but yet not put men out of work needlessly." This was the dilemma that led him to retreat from his campaign promises on strip mine controls. Through June, the Clearfield Chamber of Commerce had conducted a scare campaign, telling local residents that passage of H.B. 434 would mean an end to their jobs. In the middle of the month, the CPOPMA had sent one thousand wives and children of Clearfield County strip miners to Harrisburg on a special train, to make an appeal against control legislation. But just as important in influencing Scranton and weakening the bill was the absence of the United Mine Workers from the new regulatory campaign. Despite Yablonski's promise to return for a better bill, the measures legislators considered during the 1963 session received most of their organizational support from sportsmen's groups acting alone. The mine workers' union and other labor organizations did not make common cause with the hunters and anglers this time around. That allowed more room for the strip mining industry to exploit the divisions among potential proponents, and it eased the pressure off conservationists to speak to the particular concerns of workers. During the House debate on H.B. 434, a strip mine county assemblyman had asked former PFSC legislative chairman and state representative John Laudadio, "Which is more important—clean streams or keeping people at work?" Laudadio had bluntly and tellingly replied, "Clear water is more important."[23]

Ironically, as the struggle for control of strip mining moved to the administrative level, conservationist sportsmen demonstrated that they too could be sympathetic to the interests of the industry. The Pennsylvania Surface Mining and Reclamation Act of 1963 was administered by the Bureau of Surface Mine Reclamation, which supervised about 350 strip mine operations annually between 1964 and 1972. In addition to issuing permits, the bureau sent personnel into the field to inspect operating conditions and issue violations and "cease orders" when provisions in the law were not being followed. Within the first year of enforcement, however, inspectors developed a cooperative relationship with strippers, acknowledging their responsibility to protect the natural environment but also recognizing their responsibility for sustaining the strip mine industry. Pennsylvania Coal Mining Association director Frank Mohney was in the bureau office nearly everyday, acting as an intermediary between operators and agency officials, making enforcement of the law a bargaining process. When William Guckert took over the bureau in 1968, strippers feared the opportunities for "cooperation" with the government agency would cease, but Guckert maintained the pragmatic regulatory relationship. Like a latter-day Progressive, he was determined to eliminate favoritism and political influence from the administration of the law and to expedite its enforcement, even at the expense of conservation.[24]

This clientele approach to regulating surface mining in Pennsylvania resembled the stance the UMW International, district, and local leadership took on a whole host of environmental issues during the 1960s. Oftentimes the positions expressed by the mine workers' union and coal industry representatives on environmental regulation were indistinguishable, and in a few instances the union and coal companies purposefully cooperated to develop and voice a common position. Joined together through the National Coal Policy Conference (NCPC) in 1959, for example, coal producers, the UMW, and electric utilities worked to make air and water pollution controls more palatable to the organizations' members. When a Senate subcommittee held hearings on stationary-source (air) pollution, NCPC chair and UMW president

Tony Boyle let it be known that NCPC members opposed federal air-quality standards but could tolerate state-level regulation, which they surmised would be less stringent and more malleable.[25] Later, at the 1968 UMW Constitutional Convention, union officers noted their agreement with the alarm expressed by the coal industry over water pollution control legislation before Congress. "If permitted to stand," they explained, "many of the pollution regulations and the theories underlying them would cause a severe disruption of the coal industry and widespread unemployment for the coal miners."[26] Though the UMW had its origins in the distinct economic interests of coal company owners and coal miners, when it came to pollution controls and abatement, both viewed as a threat to the industry and therefore to jobs, the UMW and company officials thought and spoke as one.

Just as it cooperated with coal trade associations in response to air and water pollution controls, by 1964 the union was also working with coal companies to achieve a "reconciliation" with the limited state strip mine legislation that had been and was being passed. Many coal industry representatives tended to think of the control laws and their reclamation work as necessary for quieting "public clamor," and they insisted that adequate reclamation could be done voluntarily, or under state law, while federal regulation would be unnecessarily burdensome. Not surprisingly, given their stand on other environmental issues, UMW officials articulated a similar position. Speaking at the NCPC's annual meeting in March 1968, President Boyle warned against correcting problems arising from surface mining by imposing regulations that would "close down all such mining by making the extraction of coal" prohibitively expensive. The need of such regulation, he maintained, was debatable anyhow. Coal operators recognized that surface mining carried with it a deep responsibility to the public and past "abuses should and are being corrected."[27]

Yet sentiment on surface coal mining regulation at the rank-and-file level did not always match that of the leadership. Many deep miners actually favored a ban on stripping. Their opposition was a mix of concern for their homes, recognition of the jobs lost when stripping replaced underground mining, and a conservationist or environmental sensibility. Floyd County, Kentucky, deep miner Lewis Burke refused to work on strip mines because they employed only eight or nine men to the several hundred who would be employed by a deep mine, tore up the land, and destroyed the beauty of the mountains. Strip mining was his "biggest worry" because the Island Creek Coal Company owned the mineral rights to his land. "[I]f they come here to strip mine," he threatened, "there will be a war, and I mean a war, because I have no intentions of getting off *my* land." Burke claimed other miners in eastern Kentucky shared these concerns, but they were afraid to voice them for fear of reprisal from those who controlled the few jobs available. Outside of eastern Kentucky, Willie Vest, a retired deep miner in Dickenson County, Virginia, expressed a conservationist critique of stripping. "The strippers have come in here and destroyed the timber and water," he noted, "and forced the wild game to leave because there is nothing [for them] to survive on."[28]

In fact, the issue of strip mining was only one of many that divided the union leadership from the rank-and-file during the 1960s, and tensions eventually gave rise to a challenge to the corrupt Boyle regime. John Lewis had retired from the UMW presidency in 1960, passing the office on to Tom Kennedy, whose death in 1963 brought Tony Boyle, then vice president, to the union's helm. At the same time, employment in the coalfields was beginning to stabilize and rank-and-file militancy directed at coal operators as well as the unresponsive union leadership began to surface, particularly in Pennsylvania and West Virginia. Having reconsidered his own role in the UMW's failure to be an effective voice for miners, Joseph Yablonski offered to lead the activists and made a bid for the union's presidency in the 1969 national elections. Yablonski was well known to miners, having represented District 5 in one capacity or another since 1934. But Boyle was an incumbent, not reluctant to use fraud in the election, and he won 80,000 to Yablonski's 40,000 votes. The Department of Labor found so many irregularities in Boyle's campaign and the balloting that it ordered a new election for December 1972. Three weeks after the first election, however, Yablonski, his wife, and daughter were murdered on orders from UMW officials. Three officials and the daughter of one of them were later convicted of the murders, and Boyle himself was arrested for murder in September 1973.[29]

Yablonski had led District 5 in supporting the 1961 effort to enact new controls on strip mining in Pennsylvania largely for reasons of economic expediency. Following this strategic intervention, however, the UMW local, district, and International leadership aligned itself with industry officials to oppose or weaken control legislation at state and national levels. This approach ran contrary to the economic and environmental interests of deep miners, who were still the majority of the union's membership. But it was consistent with the union's general acceptance of operators' mechanization of coal mining as well as a moderate approach on environmental issues, which put economic growth before environmental quality. At least for the leadership of the UMW, the health of the strip coal industry and the preservation of strip miners' jobs took precedence over protection of deep miners jobs and the environment. As subsequent chapters will show, however, Yablonski's murder had unintended consequences for the union as a whole and its position on strip mine controls in particular. His death gave rise to an even stronger reform effort, and in 1972 union members elected Arnold Miller to the presidency. Miller was at one time a proponent of abolishing strip mining and, at least for a few years, he would chart a different course than his predecessors.

chapter four

We Feel We Have Been Forsaken

Opposition to Surface Coal Mining in Kentucky, 1954–1967

As the remarks of rank-and-file mine workers like Lewis Burke and Willie Vest (see Chapter 3) indicate, by the 1960s some opponents of surface coal mining were beginning to call for abolition rather than regulation. "Jobs have been lost, land destroyed, streams polluted, homes damaged, thousands of fish killed and children have lost their lives by drowning," complained one Pittsburgh resident in 1962. "What else has to happen," he queried, "before people realize that strip mining must be stopped?" In time, such sentiment became widespread in Appalachia, but the movement to abolish surface coal mining had its deepest roots in eastern Kentucky. In that part of the region steep "mountains" washed by heavy rains made contour stripping and auger mining especially hazardous to people, their homes, and the land, and those conditions bred militant opposition.[1]

To be sure, there was always a significant number of area residents who thought state or federal regulation alone could provide sufficient controls on the industry. Yet as early as the 1950s, residents of Letcher, Harlan, Knott, Perry, and other mountain counties began to express support for a ban. At first this abolition sentiment was diffused and uncoordinated. Organizations such as the Farm Bureau, Grange, and UMW were either not well represented in the area or unwilling to provide an institutional framework for action on the issue. In the mid-1960s, however, small farmers, active and retired deep miners, homemaker wives and mothers, as well as some middle-class professionals and business leaders banded together for the specific purpose of fighting the menace of surface coal mining. The best-known of the organizations they founded was the Appalachian Group to Save the Land and People (AGSLP), which embodied in its name the concerns of the opposition. Strip mining in the region not only threatened the land, particularly the soil and water, but also the homesteads and communities the land sustained. As AGSLP supporter and settlement school founder Alice Slone put it, "If the land was destroyed, the people were destroyed."[2]

Eastern Kentucky advocates of outlawing strip mining drew on a long American tradition of settling grievances by making formal appeals to public officials and taking legal action. Much of their activism involved circulating petitions, writing open letters and publishing personal accounts, passing resolutions, and initiating lawsuits. But strip mining opponents soon discovered that these methods were inadequate for their purpose. Drawing on another American tradition of protest, mountain residents began to rely increasingly on direct action, such as physically blocking bulldozers, sniping at strip miners, and dynamiting equipment. These militant tactics, often used by a "crowd" but sometimes employed by lone individuals, acquired their legitimacy from the impossibility of achieving justice through formal grievance processes.[3] With lives, homes, and communities in jeopardy and state legislatures, regulatory agencies, and courts corrupted by the influence of the coal industry, abolitionists took it upon themselves to make things right by nonviolent civil disobedience and calculated acts of violence (or threats of it), while also not giving up on other means. This militant activism was most intense after 1966, when Kentucky revised its regulatory legislation, and it continued through the late 1960s and up to 1972, when opponents occupied a strip mine site and made one last effort to push an abolition bill through the General Assembly. This chapter explores the events leading up to and including an important circuit court decision in early 1967, and the following chapter picks up the history of the abolition effort in the summer of that same year.

Regulatory Legislation and Early Abolition Sentiment

By the mid-1940s, a number of midwestern and Appalachian state legislatures had responded to the environmental degradation caused by surface coal mining by passing at least limited regulatory legislation. The Senate and House of Delegates of West Virginia acted first, in 1939, followed by legisla-

MAP 4. Counties of eastern Kentucky

tive assemblies in Indiana, Illinois, Pennsylvania, and Ohio. Kentucky legislators were slower to enact a regulatory measure, however, not even proposing the first control bill for the state until 1948. This first proposal was sparked by stories and editorials in the *Courier Journal*, Kentucky's leading newspaper. Yet concern during the 1948 session of the state assembly was only great enough to prompt the passage of a resolution, calling on the Legislative Research Commission to make a study and report to the 1950 legislative session. The report failed to convince legislators of the necessity of a control bill, and regulatory legislation introduced by Senator Moloney (R-Lexington) in 1952 also did not pass.[4]

Not until 1954 did Kentucky's General Assembly enact the state's first strip mine control act, perhaps the weakest of all such legislation at the time. The law required operators to put up a paltry $100 to $200 per acre to guarantee reclamation, which included little more than covering the face of exposed coal and grading spoil banks "where practicable." The legislation also established the Strip Mining Reclamation Commission, composed of the commissioner of conservation, the chief of the Department of Mines and Minerals, and a director chosen by the governor. The commission was charged with processing applications for permits, collecting and releasing bonds, making regulations, and performing inspections to correct violations. Not long after the control bill was passed, however, Governor Chandler (1955–59) abolished the agency, and in 1958 operators received a favorable court decision excluding auger mining from inspection and control. As a result of lax enforcement, by 1960 only 9 of 169 operators had permits and little was accomplished in terms of on-site reclamation or control of off-site effects. Ruined land and mudslides were visible in many eastern counties, and studies of water quality demonstrated that acid mine drainage and sedimentation from the unreclaimed mines reached critical levels, polluting surface and groundwater and harming aquatic life.[5]

In the early 1960s, Governor Combs (1959–63) reestablished the strip mining commission and his attorney general initiated legal action against operators who refused to acquire permits or abide by the criterion of the law. Concurrently, the General Assembly passed legislation adding auger mining to the strip mine law and, in 1962, it enacted another amendment replacing the reclamation commission with a Division of Strip Mining and Reclamation (DSMR). Yet many mining sites were still being abandoned or only nominally reclaimed by operators. The director of the DSMR, Henry Callis, was a former superintendent of one of the state's largest strip mining operations and reluctant to strictly apply controls. As a result, in various parts of eastern Kentucky slides of silt, mud, rocks, and trees moved down hillsides, covering farmland and ruining streams used to water livestock. After the worst in a series of three slides in 1961 and 1962, near a U.S. Steel Company strip mine in Harlan County, Wolfpen Creek was clogged with logs and mud, and a gigantic accumulation of overburden still threatened to break loose on Black Mountain. The damage done and the looming threat of greater danger forced a number of longtime residents to leave. "Fred and I have worked hard all our lives, and we've put everything we had into this place," complained one of the evacuated hollow dwellers; "I don't know what we're going to do." By 1963, even the typically apolitical Council of Southern Mountains (CSM) was using its journal to decry the increased frequency and greater severity of landslides due to surface mining, one of which threatened a settlement school. Unless the stripping was stopped or changed, an editorialist wrote, the slide "will engulf the institution—as similar operations have wrecked homes and communities elsewhere in the Appalachian South." But CSM was not so presumptuous as to suggest a solution to the problem. "Just how this wanton destruction of surface property in the Appalachian South is to be stopped,"

Landslide in front of eastern Kentucky home. (Courtesy of Mike Clark)

the op-ed piece explained, "is the responsibility of good citizens and trust-worthy officials."[6]

Other observers were skeptical that mountain people themselves could either reform or stop contour and auger mining operations in the region. Whitesburg (Letcher County) native Harry Caudill was one of the legislators who played an instrumental role in passage of the 1954 act, and he remained an important leader in efforts to improve living conditions for Appalachian people during the following decades. Yet in 1960, writing about what could be done on the strip mining issue, he described the majority of his fellow mountain residents as functionally illiterate, convinced by long experience that the coal companies always come out on top, and "unlikely to do anything of significance to help themselves." A few months later he introduced a bill to abolish surface coal mining in the state, but it received little support from his constituents and the assembly gutted the legislation. This left Caudill even more embittered about the prospects for a popular movement.[7]

In September, however, a Raymond Rash sent petitions to Governor Combs with more than a thousand signatures of Letcher County residents, calling for prohibition legislation. "Strip mining in our steep mountains," Rash explained in a cover letter, "destroys the surface for agricultural purposes, throws immense amounts of loose earth into the streams, causes mud to be carried by rain down onto our gardens and crop lands, and into our wells, and destroys the natural beauty which God has so lavishly placed in our region." Caudill was among those who put their name on the petition, as were the county judge, two ex-county judges, the tax commissioner, state senator Archie Craft, and many people who held no office at all. The secretary-treasurer of the Whitesburg Chamber of Commerce also sent a letter to the governor, following a meeting of the membership, putting the organization on record in support of the petition. As business and professional men, he wrote, "we are convinced strip-mining, as now practiced, will eventually have calamitous results on the economy of the counties directly involved and will do much long-range harm to the economy of the entire state." Using the community of Haymond as an example, the secretary-treasurer noted that most of the several hundred people who lived in the small settlement were retired coal miners living on pensions, their estate typically consisting of a home and garden, "and strip-mine operators are threatening to completely destroy these properties for all practical purposes."[8]

In 1962 Harry Caudill admitted that there was "growing interest and resentment against the stripping industry in each of the Eastern Kentucky counties," but he lamented the fact that "the interest is at the bottom of the social and economic strata." The people who were angry, the Letcher County native explained in a letter to wealthy *Courier Journal* publisher Barry Bingham, owned small parcels of land that were being destroyed for a few hundred tons of coal. Caudill was here revealing one source of the opposition movement: the broad form deed. Earlier in the late-nineteenth and twentieth centuries, agents for land companies had swept through the region buying up mineral rights, sometimes for as little as fifty cents per acre, separating the use of the surface (and tax liability) from the natural resources that might be below. To hard-pressed farmers and those actively seeking economic mobility, this exchange seemed advantageous. Written in finely printed "legalese," however, the broad form deeds often signed over the rights to "dump, store, and leave upon said land any and all muck, bone, shale, water, or other refuse," to use and pollute water courses in any manner, and to do anything "necessary and convenient" to extract subsurface minerals. These clauses eventually caused much regret on the part of the sellers' progeny when Kentucky courts interpreted them in favor of coal companies.[9]

As underground and surface mining expanded during the first half of the twentieth century, Kentucky judges were faced with a number of cases involving surface owners' complaints about shale and culm heaps dumped on their land, the pollution and diversion of subterranean and surface waters, and ground subsidence. Responding to these complaints, in 1925 the Kentucky Court of Appeals recognized the paramount right of the mineral owner to the use of the surface "in the prosecution of its business for any purpose of necessity or convenience," but denied the mineral owner the right to use the surface "oppressively, arbitrarily, wantonly, or maliciously." Although this case dealt with shaft and deep mining, it established the framework for rulings on strip mining. The court of appeals first recognized the right to strip-mine in broad form deeds in 1930, and in *Treadway v. Wilson* (1946), the justices restated this interpretation, based on the dominance of the mineral estate over the surface estate. The only limitations they imposed on the exercise of rights to the minerals, as in the earlier deep mining case, were the prohibitions against oppressive, arbitrary, wanton, or malicious conduct.[10]

Other high courts in Appalachian states delivered judgments during the postwar period that established a stricter reading of broad form deeds, even going so far as to reject the conveyance of the right to strip-mine by the deeds. In 1953, the Pennsylvania Supreme Court ruled that parties to such deeds could not possibly have meant to grant the right to strip-mine as part of the transfer of mineral rights and use of the surface. "If the grant was intended to include strip mining privileges," the justices declared, "the immunity from responsibility [spelled out in the deeds] for 'damages to the surface . . . or the failure to provide support for the overlying strata' would be meaningless because strip mining encompasses the very tearing away of the overlying strata." No surface owner (especially a farmer) would have sold their mineral rights knowing the surface could be subjected to "violence, destruction and disfiguration which inevitably attend strip or open mining." The Pennsylvania justices reaffirmed this decision a few years later, in 1961, arguing that parties to a deed separating mineral and surface rights in 1893 did not contemplate or intend that the surface of the land would be strip-mined, and owners of mineral rights were thus prohibited from using that method to extract coal.[11]

In the 1955 case *Buchanan v. Watson*, the Kentucky Court of Appeals also recognized that the original intent of parties to broad form deeds was restricted to deep or shaft mining, but it upheld the right to strip-mine by the deeds nevertheless. In 1943 Ralph and Stella Watson had purchased 20 acres of the surface of an area in Magoffin County that was included in a broad form deed from 1903, and they brought a lawsuit to prevent Elkhorn Coal from stripping their piece of the land. The trial court ruled that Elkhorn could use surface mining methods in exercising its lease of the mineral rights because it was the only economical way to mine the coal, however the company was bound to pay in full for any damages caused to the surface in the process. Elkhorn appealed, seeking the right to ruin the surface without liability, but in 1955, the higher court upheld the decision of the lower court requiring damages. Strip mining "will result in an invasion of the defendants' surface rights not anticipated by the parties to the deed," the justices argued, and for this reason "the plaintiff must pay reasonable compensation as damages to the extent he destroys . . . the defendants' interests in the surface and timber." Elkhorn petitioned for a rehearing, however, and in 1956 the court drastically revised its position on the award of damages. "The deed in this case conveyed virtually all rights necessary to carry out the mining of the coal," the justices unanimously declared in the second decision, "including a waiver of damages." Strictly following earlier court decisions, compensation was due to a surface owner only if the mining was oppressive, arbitrary, wanton, or malicious conduct (which Elkhorn's intended excavations apparently were not). Denying a waiver of damages, moreover, "would create great confusion and much hardship in a segment of an industry that can ill-afford such a blow."[12]

The rulings by the Kentucky Court of Appeals as well as weak regulatory legislation and its lax enforcement put many property owners in both the eastern and western parts of the state at the mercy of coal companies. In eastern Kentucky, mineral owners sometimes sought written consent to strip a ridge or they tried to get releases for any damage done to the surface by the mining. Just as infrequently, they offered a per lineal foot or per ton royalty to surface owners. These considerations were not required, however, and if the landowner refused to sign a release or accept a payment, the operator stripped the land anyway. More typically, surface owners worried about the lack of recourse for protecting their homes, gardens, pastures, farms, and orchards from destruction, and they lamented the impossibility of gaining any compensation. Though not suffering under the broad form deed, landowners in western Kentucky also received little protection from control legislation and court decisions dealing with surface coal mining. Area strip mine operators in that part of the state still failed to grade spoil banks and they fouled ground and surface water with acid mine drainage with impunity. This did little to endear strippers to local people. "I think there are a lot of farmers like me," explained a former president of the Hopkins County Farm Bureau, "who are completely hard against what the mines are doing."[13]

By the early 1960s, critics of strip mining from both parts of Kentucky were anxious to amend the state's regulatory legislation. But the campaign organized to do this was led largely by farm leaders, public conservation agency officials, and sportsmen who were most concerned about the situation in western Kentucky. In September of 1963, W. D. Bratcher, a Greenville attorney and leader of the Kentucky League of Sportsmen, sent out a call to various individuals and organizations for a meeting at the Game and Fish Commission Office in Frankfort. Participants in this meeting agreed to establish a committee including representatives from the Kentucky Farm Bureau, Garden Clubs of Kentucky, Business and Professional Women's Club, and Water Conservation Districts, as well as Harry Caudill, "to investigate the possibility of legislation to control strip mining in Kentucky." No follow-up occurred, however, and the Kentucky Farm Bureau and state Association of Conservation Districts took the initiative to prepare and co-sponsor a regulatory bill of their own in the next legislative session.[14]

Introduced by Representative John Swinford (D-Cynthiana), the Farm Bureau–Conservation Districts proposal contained a provision requiring that land capable of agricultural use before stripping be graded "to a rolling topography that may be traversed by farm machinery." This would be the first attempt by the state to legislate reclamation standards, and the standard would specifically address the parallel spoil piles created by the removal of overburden at area strip mines. A bill abolishing surface mining on precipitous terrain was also written by Harry Caudill and introduced by Representative J. O. Johnson (R-Jefferson), but the weaker control measure was more palatable to legislators and the abolition bill functioned as a foil. Both Representative Swinford and the *Courier Journal* warned that if there were not improvements in the control of stripping, the public would soon demand state or federal legislation to ban surface mining. Facing this possibility, in early March a slightly modified version of Swinford's bill was reported out by a House committee, the full House passed it on an eighty-eight to zero vote, and the Senate passed the measure on a thirty-seven to one vote, after defeating an amendment to prohibit strip mining in the proximity of cities and state parks.[15]

Cognizant of the limitations of the new control bill, in the year following its passage Governor Breathitt (1963–67) instructed members of his administration to develop regulatory standards for contour as well as area surface mining. To this end, in the spring of 1965, conservation commissioner J. O. Matlick assembled a two-person study group within the DSMR to prepare recommendations to establish a comprehensive strip mine control program. The two members of the study group, division chief Elmore Grim and assistant attorney general David Schneider, made a number of trips to Pennsylvania strip mines and gathered data from various other sources, including the state highway department. Their travels and research convinced the pair of the feasibility of reducing bench widths, to minimize the weight of spoil strippers placed down the slope and thereby prevent landslides, as well as the necessity of revegetation, also to stabilize the bench and control erosion. But before Grim and Schneider finished their study of the problem, an upsurge of grassroots protest during the summer of 1965 created a new context for promulgating regulations.[16]

Grassroots Mobilization for a Ban

Eastern Kentucky residents had been demanding a ban on surface coal mining since the late 1950s. But in the mid-1960s this sentiment increased, even as legislators reworked strip mine controls, and people began to organize. At the most basic level this organizing was a response to the expansion of strip mining on Clear Creek and Lotts Creek, and that expansion was largely a result of changes at the Tennessee Valley Authority (TVA). Congress had established the TVA in the 1930s and charged it with the dual mission of providing cheap power as well as conserving natural resources. The agency initially accomplished both tasks with the construction of dams, which could provide hydroelectric power and control flooding on the Tennessee River. By 1953, however, TVA was utilizing most of the available hydroelectric potential of the river and its tributaries and so it turned to coal-fired generating plants to meet escalating power demands. In 1950 steam generation of electric power accounted for only 6 percent of the total power produced by the agency, but this rose to 75 percent by 1959. To continue low-cost production of power the TVA turned to cheap strip-mined coal, primarily from eastern Kentucky and east Tennessee. The agency purchased most of this coal on long-term contracts, granted through a competitive bidding process. Since the late 1950s was a buyers' market, TVA's coal-buying policy contributed to driving out of the coal industry many subterranean operations, which could not produce as efficiently as contour strip mines. By the mid-1960s, the agency was effectively setting market trends, controlling prices, shaping the development of mining technology, and practically determining the fate of various parts of Appalachia.[17]

The Tennessee Valley Authority was not unmindful of its role in the damage done by strip mining, however, and as a public agency it was more responsive to protest than private land or coal companies. In December 1961, the director of the TVA Division of Forestry Relations, Kenneth Seigworth, called a meeting to discuss various problems associated with stripping and initiate a regional study of the matter. Attending the meeting were Central States Forest Experiment Station director R. D. Lane, Kentucky conservation commissioner J. O. Matlick, his assistant Robert Montgomery, DSMR director Henry Callis, and representatives from the U.S. Forest Service. Montgomery later wrote to Seigworth thanking him and explained, "Because TVA is a federal agency, and because they are becoming the largest user of coal produced by strip mining in Kentucky, it is definitely the feeling of Mr. Matlick that TVA should assume certain responsibilities in the area of strip mine reclamation, whether the areas lie in the Tennessee Basin or other river basins." When a regional study failed to coalesce, the agency established an interdivisional task force, which produced its own summary report in November 1962. The task force admitted "that TVA's large coal purchases, as well as those of other large consumers, contribute significantly to land and water management problems outside the Valley . . . [and] like any other agency representing the citizens, [TVA] should be concerned with these problems." This could be done by supporting passage of state legislation in Tennessee, Virginia, and Alabama, which had not yet enacted control bills, as well as through a contract clause setting minimum standards for reclamation.[18]

No policy changes followed immediately after the task force made its report, but the public continued to pressure the TVA, state governments, and the industry, and in 1963 the agency announced a meeting to consider the merits of an interstate mining compact. The TVA needed to present itself as a steward of natural resources, and some officials were sincerely concerned with the plight of Appalachian people. Governors and legislators wanted to act to quiet opposition, but they were reluctant to move for fear of putting themselves at a competitive disadvantage with other states that had little or no controls on surface mining. The coal industry sought to mine coal unhindered, yet many operators realized that agreeing to weak regulation could prevent passage of strict controls or abolition. An interstate mining compact, drafted with the input of all three groups, would establish minimum regulatory standards throughout the region and possibly stave off passage of stronger regulation.

With pledges of support from state conservation commissioners and industry leaders, the TVA convened a conference in Roanoke, Virginia, in April 1964, under the auspices of the Council of State Governments (CSG). The conference was attended by 161 representatives from eighteen states, seven federal agencies, and the coal industry. It did not produce a compact, but at the closing session participants unanimously adopted a resolution calling on states to study the problem and pass adequate control legislation. Later, at the

October 1964 meeting of the Southern Governors' Conference, the governors approved a follow-up resolution to that of the Roanoke meeting, calling on the CSG "to assist representatives of the states in which strip mining takes place in exploring the possible role of interstate action, through compact and otherwise." In January 1965, representatives from TVA as well as state and federal agencies met in the Washington, D.C., office of the council and agreed that staff from the interstate organization should prepare a draft of a compact, which was completed and distributed to the delegates from the Roanoke conference in November.[19]

Back on Clear Creek and Lotts Creek, in Knott County, Kentucky, residents were skeptical of the efforts of their state government as well as those of the TVA to control strip mining, if for no other reason than they saw evidence of the failure of regulations and the agency's concern in their own backyards. In 1961 partners Richard Kelley and Bill Sturgill, owners of the Caperton Coal and Kentucky Oak Mining Companies, signed a fifteen-year contract to supply the TVA with 2 million tons of cheap coal from mines in Knott and Perry Counties, the rights to which they leased from the Kentucky River Coal Corporation. With assistance from the federal agency, Kelley and Sturgill began to experiment with large-scale surface mining equipment for the first time in eastern Kentucky, including the largest coal auger ever built, a gargantuan machine 7 feet in diameter and capable of penetrating 216 feet into a coal seam. Operating on a tight profit margin, the strip mining pair neglected to gain the permission of landowners to extract coal under their property, which they did not need in any case, and they did little to repair the damage left in their wake. But by 1965 the companies were meeting serious resistance as blasting and bulldozers shattered windows, knocked houses off their foundations, covered roads with debris, uprooted timber, and ruined croplands without compensation to owners. In May, when the strippers reached the land in Clear Creek Valley where Dan Gibson and his wife lived, this resistance intensified.[20]

Title to the land Sturgill and Kelley sought to mine near the Gibson homestead was in the name of Dan's stepson, who was then serving in Vietnam. When bulldozers began pushing trees over the property line Dan and his neighbors went up to complain about the damage to his fence as well as to warn strip mine employees against trespassing. The next morning the bulldozers actually came onto the land so Dan took a .22 automatic and a box of shells and went with Paul Ashley up to the area. The two split up and Gibson had a confrontation with an armed guard. Dan warned the guard against drawing either one of his pistols and told him within earshot of the bull-

Dan Gibson with his rifle. (Courtesy of Mike Clark)

dozer operator to get the machinery off the land, which was promptly done. Gibson then placed himself with gun in hand at his stepson's property line and told the strippers he would not let them come through. That afternoon more than a score of police officers made a number of attempts to arrest Dan but he refused to be taken. Toward evening, strip mining opponent Eldon Davidson went over to where the police were milling around and told them that Gibson would give up his gun and go off the hill if he could get a promise the strippers would not cross onto his stepson's land. An officer called over to Sturgill in Hazard, Sturgill made the promise, and Dan's neighbors reassured him they would keep the bulldozers off, so he voluntarily gave himself up for arrest and was incarcerated in the Hindman jail. Soon after Gibson was incarcerated, the jail was surrounded by armed men demanding his release, the charges against him were dropped, and he was freed. The next morning, when the bulldozers returned to work, they were met by what company lawyers described as "a big gang of outlaws," nearly all of them elderly, some of them women, and a number of them armed, who placed themselves at the property line and refused to allow the equipment to pass. And it never did.[21]

A few weeks after the standoff on Clear Creek, on June 1, eighty people assembled in Hindman, the Knott County seat, to discuss ways of preventing destruction of their homes and farms. They declared their willingness to resort to sit-ins, lie-ins, and even guns to keep strippers off their land. They also brought with them petitions signed by eight hundred residents of Knott and Perry Counties. Land sharks had cheated their forefathers out of their mineral rights, the petitions read, paying fifty cents to $1 per acre. Now strip miners were boring, ripping, and tearing away at the topsoil to get at the coal underneath, and in the process rolling stones, boulders, trees, and dirt down onto private property, homes, and land. In addition, there was "the inevitable acid water which follows the auger and strip mining to complete the cycle of destruction by killing fish, trees, grass, anything it touches" and seeping into water supplies. The people of Appalachia had always been willing to send volunteers off to war to protect freedom and basic human rights guaranteed by the Constitution of the United States, the petitions went on, but strip miners were undermining those guarantees. "We feel we have been forsaken," they explained, "that we have no rights when a county sheriff can order a man off his own property and tell him he is trespassing; that he will be jailed if he doesn't readily comply."[22]

The next week an even larger number of local residents met and formed the Appalachian Group to Save the Land and People, which was dedicated to stopping completely all contour and auger mining in eastern Kentucky. The new organization was 125 strong and composed largely of landowners, many of whom were active or retired deep miners and farmers, but with a few school-teachers and merchants thrown in for good measure. At their first meeting the strip mining opponents chose schoolteacher Leroy Martin as their chairman, Perry Commonwealth attorney Tolbert Combs as co-chairman from that county, and Jenkins high school principal Eldon Davidson as co-chairman from Letcher County. Martin's interest in the issue included a direct threat to his home by strip mining and a lack of faith in both regulatory legislation and the courts to provide him either protection against destruction or compensation for damages. "They have got all the laws set up for the operator against the man who owns the land," he said, "and we don't have a chance." Davidson questioned the idea that strip mining was essential to the area economy. "I would dispute the claim that we are any longer dependent for our economy upon strip or auger mines," he said; "I don't think there is anybody in this state who can justify the destruction of these mountains." On Wednesday of the following week, June 16, Judge George Wooten led a small delegation to the second AGSLP meeting at Carr Creek High School and the group elected him co-chairman for Leslie County. Wooten not only rejected the argument that coal surface mining was necessary for the economy but also insisted that it did more harm than good. The strip mine industry was taking the wealth and destroying the land of the mountains, he argued, leaving nothing but a lot of poverty.[23]

One of the first official actions of the new group was a fifty-car motorcade to the state capital in Frankfort, to meet with the governor. Breathitt had initially refused to meet with group members, claiming the state's reclamation law would prevent "undue damage," and when the activists arrived they were told that he was in a conference and would probably be too busy to see them personally. When a reporter for the *New York Times* pointed out how badly this would reflect on Breathitt's administration, however, the governor arranged to see ten AGSLP leaders in his office, and then he and assistants met the entire delegation for three hours in a health department auditorium. The activists booed natural resources commissioner J. O. Matlick as he told them about enforcement of the strip mine law in Knott County, and they presented Breathitt with petitions bearing three thousand signatures claiming that strip mining operations in the mountains were "ruining our farms and fields and streams." The governor made only a vague response, announcing that "if [the 1964 law] proves to be insufficient, then I intend to ask the Legislature for a law that will do the job." At the end of the day he also pledged to initiate a study of current regulatory policy and tour some Knott County strip mines.[24]

The next evening, Breathitt told a Lexington symposium on strip mining that operators would have to protect the people of eastern Kentucky themselves or he would ask legislators to completely outlaw stripping in the mountains. But prohibiting coal surface mining would have serious consequences on the coal region, the governor acknowledged, so "we must make sure that we have explored every possible avenue of reclamation and enforcement" before taking such a drastic step. On an inspection of Knott County a few days later, Breathitt surveyed some of the damage done by strip mining for himself. This included a visit to the Ritchie cemetery, where strippers had disinterred coffins years before, and a stop to talk with Claudia Hall, whose family's 40-acre farm was threatened by debris from nearby mines. At one point in the tour Baptist preacher Frank Fugate explained to Breathitt how he and his neighbors had temporarily stopped the strip mining on his property, until the Hindman court had issued a restraining order against such protests. "We didn't want to go to violence," he said, "but we knew we were ready for it." The governor was reportedly moved by what he saw and heard, although judging by his comments he might have been simply shocked by the level of tension in the community and the determination of its residents to take matters into their own hands. "We do not intend to permit [the strip mine] in-

Bessie Smith and Madge Ashley blocking coal trucks coming off a strip mine in eastern Kentucky. (Courtesy of Phil Primack)

dustry, or any other industry," he exclaimed in good conservationist form, "to destroy the beauty of Kentucky's countryside or the usefulness of its earth for future generations." But Breathitt also took note of the militant abolitionist sentiment brewing in the mountains. "Unless we solve these problems," he warned, "strip and auger mining in Kentucky and the nation will be seriously threatened."[25]

At the end of the inspection tour the governor announced that his administration would take a number of steps to end the damage caused by contour and auger mining in eastern Kentucky. First, he would ask the state attorney general to intervene as a friend of the court in cases attacking the prevailing interpretation of broad form deeds. This would include a Knott County case brought by the Kentucky Oak Mining Company to prevent interference with its strip mining operations. Second, Breathitt pledged to call for an early conference with the Tennessee Valley Authority to discuss the agency's coal-purchasing policies. He wanted the TVA to pay coal operators an additional amount to ensure they would reclaim the mountainsides where they mined coal on contract. Finally, the governor planned to ask the state department of natural resources to prepare regulations controlling strip mining on steep slopes (which it was already doing), and he told Grim and Schneider that he would support anything they drafted.[26]

In mid-July, assistant attorney general David Schneider requested a Knott County Circuit Court to make the state a co-defendant in the Kentucky Oak Mining case against nineteen landowners. His application for intervention included a plea to prohibit strip or auger mining under broad form deeds and where such operations result in damage or peril to members of the public, streams, roads, and other public property. To interpret the broad form deeds "so as to give the mineral owner a complete license to destroy large portions of the entire surface," Schneider argued, "would in fact make surface ownership a mere nullity, depriving the surface owner of the use and enjoyment of his land without any relief or remedy at law." The assistant attorney general justified the state's petition for intervention on the grounds that the court's decision in the particular case "will necessarily involve numerous other landowners holding title under similar circumstances," and thereby would vitally affect the rights and interests of the state in its duty to protect public property and natural resources. But Judge Don Ward ruled that the state could intervene as a co-defendant only in points involving damage to public property, such as streams or roads, and he denied Schneider's petition.[27]

Later in July, while he was attending the National Governors Conference, Breathitt called for a meeting of governors to organize joint action in the control of strip mining. He circulated a "statement of principle" among the chief executives urging Congress to enact minimum standards for reclamation of surface-mined land, encouraging inter-state compacts for the control of strip mining, and leaving regulation and enforcement in the hands of the affected states so long as they enforce the federal standards. "I have always believed, and continue to believe, that state problems should be handled by states," he said, but "in recent years the practice of open-pit or strip mining has confronted Kentucky with a problem of such scope and gravity that it cannot adequately be handled by the state alone." Pennsylvania governor William Scranton reacted cautiously to the proposal, favoring an interstate compact but hesitating to support congressional action that might result in regulations weaker than those in his own state. Other governors, seeking to avoid federal regulations for fear they would be stronger than what their states had passed or might pass, forced a compromise. The revised "statement of principle," drafted by Breathitt and Scranton, pledged support for interstate compacts designed to achieve uniform standards for regulating strip mining, called upon the federal government to "set an example" by requiring reclamation practices from its coal suppliers, and urged Congress to authorize a "strip mining reclamation fund" to supplement the financing of the Appalachian Development Act and other regional programs in dealing with abandoned mine lands. Forty-four governors signed the statement and it was sent to the president and members of Congress.[28]

Some of the work to reform TVA policy came to fruition too. Responding to a news article in the *New York Times*, Tennessee Valley Authority board member Aubrey Wagner explained in a letter to the editor that he thought states should have the responsibility for controlling surface coal mining but, if they failed to act quickly, "Federal regulations must be the answer." He also agreed that the cost of reclamation should be included in the price of the coal sold to the TVA. At the end of the month, the agency's board approved a contract provision, "Surface Mining Reclamation and Conservation Requirements," which was in line with the "statement of principle" Breathitt was then circulating at the National Governors Conference. When the agency announced invitations for bids on a contract in August, it included stipulations requiring any supplier to cover stripped coal faces, seal breakthroughs into abandoned deep mines, avoid the deposit of overburden and spoil in streams and on roads, control water runoff, as well as regrade and revegetate stripped lands. Wagner cautioned, however, that "the contract provisions for reclamation can not and do not provide an answer to the over-all problems of strip-mine reclamation," and he called again for "strong, well-enforced legislation" by the coal producing states.[29]

Also in August, David Schneider and Elmore Grim announced their proposals for regulating contour and auger mining, limiting the height of highwalls, restricting bench width, and setting specific requirements for revegetation. Additionally, they outlined regulations which would require the grading of strip mine sites in western Kentucky so that they could be traversed by farm machinery. In response to the proposals, members of the coal industry organized a protest. At the end of the month, hundreds of independent truckers drove to Frankfort in their own motorcade to see the governor, to make the claim that strip mining was the only viable industry providing jobs and tax revenues in eastern Kentucky. After hearings on the regulations were completed in November, however, Breathitt signed an emergency proclamation putting them into immediate effect. Shortly after that he instructed the department of natural resources and office of the attorney general to draft comprehensive strip mining legislation for the 1966 General Assembly, a task that fell largely to David Schneider. "In essence," Schneider later explained, "the legislation elaborated on and gave statutory blessings to the administrative regulations."[30]

The AGSLP went into action at the end of November as well, when sixty-one-year-old Ollie Combs attempted to prevent her homestead in Honey Gap (Knott County) from being destroyed by the Caperton Coal Company. Combs had telephoned Dan Gibson asking him what could be done to protect her four-room house and land, as well as that of the other eleven Combs heirs, and Gibson advised her to join the Appalachian Group. She paid the $1 for annual dues and, a few days later, Combs called Gibson again to tell him that a bulldozer had started moving overburden. Dan went over to Honey Gap with four other men, finding many other local people already there, and the group ran the strippers off the land by threat of armed force. Questioned about the possibility of taking the dispute to court, Sam Hollifield—a defendant in the Knott County court case, a husband of one of the Combs women, and the spokesperson for the Honey Gap landowners—gave a vague answer about having it in court before, "but the judge won't make a decision." AGSLP activists guarded the land for a few days, but when they left the strip miners returned. Facing them with only the aid of two of her sons this time, Ollie Combs sat down in front of a bulldozer. She and her sons were arrested and declared in contempt of court, for violating the injunction against hindering strip operations in Knott County, and Judge Ward sentenced them to twenty hours in jail. When a picture of Combs eating Thanksgiving dinner behind bars made the *Courier Journal,* Governor Breathitt declared his support for her and all other Kentuckians whose homes and farms were threatened with destruction by strip and auger mining. Breathitt urged all citizens of the state to obey the law but noted "history has sometimes shown that unyielding insistence upon the enforcement of legal rights by the rich and powerful against the humble people of a community is not always the quickest course of action." He then instructed the public safety commissioner that state police were not to be used for the enforcement of civil processes in the courts without permission from the governor and also revoked Caperton's permit to mine the Combs land.[31]

Coming as it did in the wake of violence and threats of violence by others in her community, the nonviolent civil disobedience performed by Ollie Combs had an impact on Kentucky legislators during the next session of the General Assembly. But it did not convince them to support abolition or reinterpretation of the broad form deed. When the legislature opened in early January Governor Breathitt told the representatives and senators that strip mining regulation was the foremost issue requiring consideration. "Experience over the past two years has shown, despite the most conscientious and energetic efforts at enforcement," he said in his opening address, that "our present laws are not adequate to protect the people or their land and to deal fairly with the coal industry." This appropriated the rhetoric of AGSLP and yet left room for conciliation with strip operators and their legislative allies. Likewise, the regulatory bill that Breathitt had a central Kentucky representative introduce was not presented as part of the administration's own program but rather under sponsorship of the Kentucky Association of Soil Conservation Districts, a tactic which made it easier for the governor to accommodate coal interests by agreeing to changes. The bill was intended to write into law the regulations Breathitt promulgated in November, to further improve grading standards for area strip mining in western Kentucky, and to provide the DSMR with greater powers of enforcement. Strip mine lobbyists visited the governor even before the proposal reached committee, however, and he removed the abandoned mines land fund, reduced the acreage fee from $50 to $25, exempted acreage devoted to access roads from the fee requirement, and allowed for reclamation plans to be designed without consultation with the local soil conservation district.[32]

While the Breathitt administration and the coal industry were mustering their forces to pass a bill that would meet both of their interests, Harry Caudill was in communication with Oz Johnson, a member of the House from Louisville, to convince him of the need to introduce an abolition bill. Prohibition of contour and auger mining in eastern Kentucky and "total reclamation" in the western part of the state were the "only real and adequate response[s]," he wrote Johnson, and "it would be a wonderful thing if you would resurrect your old bill from the 1964 session and introduce it in the current session." Caudill claimed that former governor Bert Combs had told him by telephone that "prohibition is the answer," and perhaps this was indicative that such a measure would pick up some support. Using a tactic that middle-class leaders would employ in other states and at the federal level, he also explained to Johnson that an abolition bill "would certainly make the Governor's bill much more likely of speedy passage." A bill to ban strip mining could be effective as an unacceptable alternative to pressure legislators to pass a strict regulatory bill, even though most of the grassroots advocates of abolition were seeking no compromise. In mid-January Johnson wrote back to say he was having Legislative Research prepare a bill that would outlaw strip mining in eastern Kentucky, but he was not sure if he would introduce it "because of the apparent effort that the administration is making to pass their bill." In the end, Johnson did not bring an abolition proposal before the House.[33]

To promote his own measure, Governor Breathitt rented aircraft to take members of the assembly on tours of strip mines in both the western and eastern parts of the state, and he staged a three-day "Pre-Legislative Conference" as well. The conference brought three officials from Pennsylvania— the coal industry's representative on the state's reclamation board, the director of the conservation and reclamation agency, and the president of a coal company—to testify on the opportunities for making a profit under Pennsylvania strip controls. At the same time, J. O. Matlick attended dozens of Farm Bureau meetings to drum up support for the regulatory bill, and various other organizations began to announce public endorsements. The president of a local of the Southern Labor Union, a renegade miners union in eastern Kentucky, reported that the 240 members he represented supported the measure. Strip mining was detrimental to the economy of the area, he said, "it produces dirt-cheap coal at the cost of ruined land and streams, puts underground miners out of work, and has put thousands of people on public assistance." The nine district vice presidents of the Kentucky League of Sportsmen, representing 36,000 hunters and anglers, also unanimously supported the bill in the interest of protecting the state's "fish, game and forests." And the Kentucky Civil Liberties Union backed the measure, arguing that strip mining practices "violate the basic rights of the property owners and the owners of adjoining property."[34]

In mid-January both proponents and opponents of the Breathitt bill, as well as advocates of prohibition, had the opportunity to voice their positions in three days of joint hearings. The main witness for the coal industry was E. W. Phelps, vice president of Peabody Coal, a company that DSMR inspectors considered to be one of the most uncooperative. He was followed by representatives of utilities who stated that a rise in the price of coal as a consequence of added reclamation costs might force them to look elsewhere for their fuel purchases. This not-so subtle effort at blackmail left legislators more receptive to hearing the proponents for controls. Representing the Kentucky Farm Bureau, E. W. Keller said his organization was concerned about long-term economic and social effects of strip mining. "If we permit the agricultural potential of those counties [facing large-scale strip mining] to be destroyed, we will destroy their ability to finance their schools, county roads and many other functions of a county government. The tax base will be gone." Others testifying in support of the regulatory bill included Harry Caudill, Elmore Grim, the state conservation chair of the Kentucky Federation of Women's Clubs, the past president of the Garden Club of Kentucky, nature writer Wendell Berry, the president of the Kentucky League of Sportsmen, Leslie County judge George Wooten, and Governor Breathitt. Caudill was one of the few to also advocate prohibition of mountainside stripping. "The welfare of the state and its citizenry," he said, "should be placed above that of any little group of exploiters." Breathitt echoed this sentiment for a different purpose, calling on legislators to choose "between the resources of soil, of water and of beauty with which a good God has endowed this beautiful commonwealth, or the exploitation of these resources for the profit of a comparatively small number of powerful corporations, many of them absentee-owned, who would claim the privilege of exhausting our children's inheritance to provide cheap fuel." But the most powerful testimony came from outright opponents of stripping, particularly Bige Ritchie, whose family graveyard had been dug up by strippers in 1959, and Ollie Combs, who demanded something be done to stop coal surface miners from destroying homes, the land, timber, and streams.[35]

After the hearings the administration's control bill received a favorable committee report and passed the House on an eighty-three to ten vote. The Senate passed the measure on a thirty-six to two vote, with senators from Pikeville and Tutor Key, both coal mine areas, voting no. Breathitt signed the bill a couple of days later at the home of J. O. Matlick, who had suffered a heart attack following a confrontation with coal industry representatives. In addition to making Kentucky the first state to agree to the interstate mining compact drafted by the CSG, the regulatory legislation specified grading "to approximate original contour" after area strip mining operations, proscribed bench widths and maximum slope angles for contour mining, and detailed revegetation requirements for all sites once mining was completed. The law also created a Division of Reclamation, granting it the powers to levy fines between $100 and $1,000 per day for each violation, to suspend permits in case of serious violations, and to revoke the permits of repeat violators. The entire budget of the division would come from permit fees, however, meaning the only way the regulatory agency could raise the funds to hire and train the personnel needed to enforce the act would be by issuing more permits. The law was weakest in terms of its provisions for enforcement, providing for only a handful of inspectors to make thousands of inspections on tens of thousands of acres of land.[36]

Besides imperiling the effectiveness of the control bill with weak enforcement provisions, state legislators and the governor also failed to move the "Widow Combs" bill through the legislative process. Introduced in tandem with the regulatory proposal, the measure would have prohibited the holder of a broad form deed from employing any mining method not in common use at the time the deed was signed, and thereby address one of the key concerns of the militant activists in eastern Kentucky. Mineral owners' trespass on land without surface owner consent and the destruction of homes and farms with little or no compensation were primary grievances of the abolition movement in the mountains. But Breathitt did not designate the bill as requiring "immediate consideration" in his address to the General Assembly, and the proposal generated much more vehement opposition for fear of the effect it could have on the availability and price of coal. Coal representatives claimed that the regulatory bill alone would put them out of business by adding $1.50 a ton to production costs (state officials estimated five to ten cents per ton). They also made an effective appeal to legislators that abolition of the broad form deed would sink the strip mine industry and dramatically impact railroads and steel, among others. As a result, at the end of January the House tabled the "Widow Combs" bill on a sixty-nine to eleven vote, killing its chances of passage.[37]

Strip mining opponents in eastern Kentucky were relatively inactive for the rest of the year, but in the early part of 1967 the legal efforts of the AGSLP in Knott County brought a state court decision that seemed to limit the scope of broad form deeds. Appalachian Group members had been making appeals at meetings for donations to establish a legal defense fund, and, from women chipping in money they had set aside for other purposes, bake sales, and sewing bees, the fund soon amounted to $3,000. The AGSLP then filed suit for a declaratory judgment against the Kentucky Oak Mining Company on behalf of Leroy Martin. Martin had been a schoolteacher at Cordova High School, and he and his wife owned the surface rights to 10 acres of bottomland in Knott County, on which they had built a house and outbuildings and planted a garden and fruit trees. Kentucky Oak owned the mineral estate on the property and the timbered hill behind them by a broad form deed dating back to 1905. The ridge had already been deep mined in the 1930s and 1940, but the outcrop remained. The Martins argued that strip mining this outcrop would send mud and rubble down the hill, ruining their "improvements," and therefore it should be disallowed. "All we want to do," said Leroy in reference to the larger significance of his case, "is save our homes."[38]

In January 1967, Knott County Circuit Court judge John Chris Cornett ruled that Kentucky Oak had the right to use contour or auger mining methods but the right to use did not convey the right to destroy. "If such were the case," he argued, "no one could safely conserve or improve or build upon the surface lands wherein the coal has been . . . conveyed [separately to others] by the broad form deed." In the case of strip mining when mineral and surface rights had been separated, Cornett decided, the operator would be liable to the surface owners for damages to the estate in land, including surface, timber, vegetation, water supply, fences, buildings or other improvements. Both sides appealed this ruling, however—the Martins maintaining that the judgment was erroneous in allowing surface mining at all, and Kentucky Oak contending that the judgment was in error in imposing upon the mineral owner the obligation to pay damages. Governor Breathitt pledged to continue giving legal assistance to the Martins, assigning his close advisor Edward Prichard and Assistant Attorney General Schneider to intervene on behalf of the state, and the case made its way to the Kentucky Court of Appeals.[39]

chapter five

We Will Stop the Bulldozers

Opposition to Surface Coal Mining in Kentucky, 1967–1972

Passage of the 1966 control legislation and the 1967 court victory by no means solved the problems caused by strip mining in eastern Kentucky, and they only temporarily quieted the opposition. The number of permitted acres actually increased between 1965 and 1967 and, as the industry expanded, the destruction of homes, wells, gardens, orchards, and farms continued while the movement to abolish surface coal mining grew. In Knott and Letcher Counties, people collected more than four hundred signatures on petitions to congressional representative Carl Perkins, claiming that the majority of residents in Knott, Letcher, Perry, Harlan, and Pike Counties wanted strip mining stopped and requesting him to bring secretary of interior Stewart Udall to the area to see the damage. In Harlan, residents on Little Creek also began circulating petitions to Governor Breathitt, explaining that stripping on steep slopes endangered private property and public safety and asking him to revoke the permits of strip operators and to refuse to issue new permits.[1]

Much of the new opposition to strip mining in eastern Kentucky began as local, isolated protests, primarily concerned with the devastation on a nearby ridge and organized through informal networks of neighbors and relations. But it was usually not long before someone contacted the Appalachian Group to Save the Land and People and the local opposition then became part of a larger, more structured campaign to stop strip mining. Mountain residents also received assistance from antipoverty workers, originally part of a program sponsored by the Council of Southern Mountains but later associated with the federal Office of Economic Opportunity (OEO). These activists, some natives and others outsiders, were forced to the realization that painting one-room schoolhouses and other service work was not sufficient to tackle the massive structural inequalities in the region. By 1967, they were playing an important part in encouraging local people to help themselves by taking their rightful places on community and county planning boards, starting cooperatives, and organizing against strip mine operators. Toward the end of the decade, however, both AGSLP and the federal antipoverty programs were in decline, for reasons that are explained in this chapter. Their role in coordinating opposition to surface coal mining was filled by a new organization, Save Our Kentucky, which became an important component in growing regional and national movements to outlaw stripping.

AGSLP and Antipoverty Workers

The Appalachian Group responded to the renewed activism in the second half of the 1960s by sending out organizers and producing propaganda. In the early part of July 1967, Knott County member Mart Shepard attended a meeting of seventy-five Harlan County residents to explain the Appalachian Group's position on surface coal mining and possibly form a new chapter. AGSLP felt that it was unconstitutional for private property and homes to be damaged and destroyed by coal companies operating under broad form deeds, he said, but it was not fighting the coal industry. The group would actually like to see more deep mines developed since they provided jobs without destroying the land. But "if an industry has to operate at the expense of poor people," Shepard argued, "it ought to be outlawed." Other speakers at the meeting included several residents of the Jones Creek area on Clover Fork and the Little Creek area on Martins Fork, where the petitions to Governor Breathitt first began to circulate. Some families in those areas had already been forced from their homes as a result of stripping and, with more mud, rocks, logs, and stumps accumulating on hillsides and in streams, the local residents also wanted an end to strip mining. After Shepard affirmed the concerns of the Harlan County residents they joined the Appalachian Group.[2]

AGSLP also spelled out its understanding of the strip mine problem and the organization's goals in a basic pamphlet, "Strip Mining—Questions and

Answers." The purpose of AGSLP, it said, was "to see strip and auger mining outlawed by the federal or state government." Surface coal mining was not a critical part of the economy of eastern Kentucky, the pamphlet declared, because "a very large strip-mining operation bringing millions of dollars of profit to the operator can be run with a handful of men," and oftentimes these workers were not even from Kentucky. In fact, it cost state taxpayers a dollar for every ton of coal strip-mined, figuring in reclamation, reforestation, road repairs, flood control projects, and water pollution. This situation was made worse by the lack of a severance tax on minerals, allowing operators to extract tons of coal without returning any of the profits they realized to local communities. But strip-mined land could not be reclaimed in any case, the pamphlet went on to explain, because the terrain was too steep, the area received too much rain, and seedlings washed down spoil banks. "Every year we have more mud slides," it said, "and the situation has gotten worse, not better, since the new strip mine control law was passed in 1966." With active chapters in five counties and members in more than ten others, the group was "prepared to stand [bulldozers] off by every lawful means."[3]

Despite this resolute declaration at the end, however, within AGSLP there was apparently some debate about tactics, part of which is evident in the working paper "The Appalachian Group to Save the Land and People: What Kind of Action?" Using the courts and engaging in political action were two possibilities for stopping surface coal mining, the paper noted, but the coal companies influenced judges and "elected officials to do what they want them to do because they have money." People could organize and help themselves, but the question—considering the power of the operators—was what should they do? "[Opponents] can picket those who support strip mining at the expense of the people," the paper explained, which draws the attention of other people to the problem. "They can block the strip mine bulldozers or trucks by standing in front of them," which gives the operators bad publicity and builds public support for outlawing stripping. And opponents could "(quietly) use guns and dynamite to take out strip-mine equipment," causing financial hardship to coal companies. By the summer of 1967, in fact, sniping at bulldozer operators and sabotage with explosives had been employed a number of times by opponents of stripping, but the paper claimed that there were only rumors these things had been done.[4]

Probably the first use of industrial sabotage for the purpose of stopping strip mining was in April 1967, when a diesel-powered shovel belonging to Kentucky River Coal was dynamited in Knott County. In June, another diesel shovel valued at $50,000 was blown up at a Kentucky Oak stripping operation nearby. That same month strip mining opponents exchanged gunfire with workers and dynamited a grader at the Tarr Heel Coal Company mine on Lost Creek (Perry County). "They ran my men off," said Harold Sigmon, the head of the company, but "there was some shooting on both sides after that." In early August, saboteurs used carbon nitrate to destroy a $90,000 auger, a D-9 bulldozer valued at $84,000, two trucks, two drills, and a welder at the same site.[5] And around that time, some Knott County activists also formed a new "conservation group," as Buck Maggard put it, called the "Mountaintop Gun Club." The club assisted surface owners who feared that a strip operation would move in on their land by leasing the surface for $1 and setting up a firing range. Yet despite the apparent inclination of some opponents of strip mining to engage in violent acts as well as nonviolent civil disobedience, both of which continued in 1968, "What Kind of Action?" rejected these approaches, concluding that "strip mining can be stopped—legally." This might suggest that the working paper was written or heavily influenced by one or more antipoverty activists, whom some mountain residents later chided for attempting to steer advocates of a ban on strip mining toward legal, formal means for resolving grievances.[6]

Antipoverty workers were not new to Appalachia, but a new brand began entering the region as part of the Appalachian Volunteer program in 1963. The "AVs," as they were called, were college students sponsored by the Council of Southern Mountains to perform weekend and summer renovation service projects, such as repairing one-room schoolhouses. In 1964 the AVs began to receive federal funding through the Office of Economic Opportunity and the program entered a contractual relationship with Volunteers in Service to America (VISTA). By the summer of 1965, there was a noticeable shift in AV work, which then included VISTAs too, as nearly five hundred antipoverty workers began to arrange community meetings and organize people around common problems, including surface coal mining. The next spring, the OEO supplied the organization with federal funds to expand operations into West Virginia, Virginia, and Tennessee, largely because of its success during the previous summer in leading "resident participation." In May 1966, however, the AVs separated from CSM as the result of a disagreement over the involvement of volunteers in activism and after the firing of three staff members. As a nonprofit corporation the council was reluctant to engage "political" issues, as opposed to "social" issues, and its history as a respectable social uplift organization was in tension with the community organizing that AVs and VISTAs were doing.[7]

For antipoverty workers, the summer of 1966 was full of efforts to bring

Sabotaged mine site. (Courtesy of Mike Clark)

about "maximum feasible participation" of the poor in the transformation of their own situation. It was not until the end of the summer, however, that they made the shift from a service to a community organizing orientation in full. In late August, activists brought residents from Kentucky, West Virginia, and Virginia to Washington, D.C., as part of a program called "Appalachia Speaks," to discuss local and regional problems with federal officials. The whole contingent stayed at a private high school, the Hawthorne School, and on the night of the last day nearly 150 people gathered there for a community meeting. "The thing that really came through at that meeting was strip mining," recalled AV Joe Mulloy, "how that was ruining the land, people had no recourse, couldn't stop it, couldn't do anything, [and this] caused a real introspection within the AVS themselves." The AVS and VISTAS were also encouraged to move "from self help to sedition" when they were joined by activists working for the Southern Conference Education Fund (SCEF), whose executive director, Carl Braden, was avowedly political and radical and hoped to "build a strong enough movement so that we can be of real help in abolishing strip mining in Eastern Kentucky." By November, Braden had placed organizers in a number of communities, including Hemp Hill, where residents were petitioning the Kentucky Department of Natural Resources (KDNR) to exempt their area from a strip mining permit.[8]

During the summer of 1967, AVS and VISTAS were heavily involved in helping AGSLP to spread and grow. The "Strip Mining Bulletin" the Appalachian Group distributed on a monthly basis was published in Bristol, Tennessee, by the AVS and it reached nearly four thousand people. Appalachian Volunteers also put out AGSLP press releases; researched land and mineral ownership, employment, permits, and reclamation; and developed plans for a lobbying campaign to influence state and federal legislative action. In July, antipoverty workers helped arrange transportation for two hundred strip mining activists from eleven counties to picket a reclamation symposium in Owensboro, which included KDNR commissioner J. O. Matlick, Kentucky Coal Association president Fred Bullard, and Governor Breathitt. Though not originally on the program, AGSLP was donated twenty minutes time by the state after a scheduled speaker announced she could not attend, and Eldon Davidson went inside to speak for the group. Outside in the parking lot antipoverty workers Michael Kline and Guy Carawan played guitars and led everyone else in singing "Strip Away" and other songs. The protestors also met briefly with the governor and distributed "Why We Come to Owensboro: A Report to Governor Breathitt." "We must point out to you that a bulldozer moves much faster than courts and legislatures," it explained, "so, while we wait for action by the state, we are prepared to protect our land by whatever means are necessary." The Appalachian Group was opposed to violence, the report made clear, "but we cannot help noticing the increasing number of Eastern Kentuckians who are telling us that the only language the strip-mine operators understand is the language of bullets. . . . Think about what steps you would be willing to take to protect your home, if the law could not or would not protect it for you."[9]

AVS and VISTAS were also involved in the July showdown in Pike County between Island Creek farmer Jink Ray and the Puritan Coal Company. Pike County was one of the nation's richest coal counties, but it was dominated by independent coal operators, and this was reflected in the poverty of its people. Coal mines produced $65 million worth of coal in 1966, but the county could raise less than a quarter of the $4 million it needed to operate its schools. Nearly half of its residents subsisted on incomes below the poverty level. On top of these indignities, strip coal operators exercised little restraint in destroying the surface when extracting coal, as was their legal right. When community leaders began to meet to fashion a response, AV Joe Mulloy put them in contact with AGSLP, an Island Creek chapter was formed, and by June it had forty paid and active members. In early July, when a Puritan Coal bulldozer crossed over Jink Ray's property line to prepare a bench for an auger operation, he and twenty-four neighbors forced the bulldozer back to the bench. Ray Johnson, the head of the company, had offered Jink Ray $2,500 for access privileges, but he had refused to take it. "If they go across me," Ray said, he would have to leave the land he owned since 1921, and then the company "can go to the head of this creek . . . [and] take a fine stand of young white poplars worth thousands of dollars, a whole lot more than they have offered to pay me." Activists vowed to defend the land "as long as they can and in any manner that they can," and they did so even after Judge George O. Bertram issued an injunction against interfering with the mining.[10]

During the second week of July, Johnson claimed that someone set off a charge of dynamite near his machinery, although the AGSLP chapter officially disavowed violence in any form. "I can say without reservation," explained Island Creek chapter vice president Marvin Thorn, "that we're not going to do anything illegal. No violence." But the failure of the 1966 law as well as Judge Bertram's ruling against any interference with the operation had put activists in an agitated state. "Some of our men are mean and hot-headed," said Island Creek general store owner and AGSLP member Arthur Akers, "and they can pick off a bulldozer without batting an eye." Retired deep miner Bill Fields explained, "We mean to stop them, one way or another. We'll use all the good means first. Then we'll use the bad ones." When reporters prompted him if that meant violence, he elaborated: "If they try to run that bulldozer in here another foot, there'll be blood spilled on the mountain."[11]

Although no violence occurred, when Puritan tried to resume its operations on the morning of July 18, local residents gathered and stopped the work. Don Branham and Carl West stood in front of a bulldozer coming near Ray's property line, pushing tons of boulders toward them. Branham mounted one of the boulders and waved to the bulldozer operator to go back and he did. Ray had been named personally in the injunction and he held back under cover of trees with other neighbors during the showdown. Shortly after Branham and West chased the bulldozer off, Pike County sheriff Perry Justice arrived on the scene to pass out copies of Bertram's order and put the landowners on notice that they were subject to arrest for contempt. But Governor Breathitt had already taken action of his own, in response to an urgent telegram from Ray pleading for help. He ordered a temporary suspension of operations on the grounds that the stripping would ruin Ray's property. The suspension was based on an infrequently used provision in the 1966 law that allowed for excluding areas from surface mining if the operation threatened private or public property. Conditions at the Island Creek mine, said Breathitt's order, "show that the approved reclamation plan cannot be carried out unless additional measures are taken to eliminate damage to the public and to adjacent property owners from soil erosion, water pollution, and hazards and dangers to life and property."[12]

Days later, reclamation director Elmore Grim surveyed the Puritan strip mine site and announced that two key sections of Jink Ray's property—the areas where he and neighbors had blocked the path of bulldozers—would have to be deleted from the permit initially granted by his agency because their grades were steeper than the 33-degree maximum allowed by regulations. Grim also announced that Ray Johnson would probably be issued a permit the next week to strip mine 20 acres on the Right Fork of Island Creek, owned by two land speculators seeking the ten cents per ton royalty that Puritan would pay. On August 1, the Division of Reclamation director permanently canceled Johnson's permit for Jink Ray's property, citing the provision in state law Breathitt had used to temporarily halt the operation. Puritan appealed Grim's decision to the Reclamation Commission, however, and at the initial hearing on August 30 the company asked for more time to prepare their appeal. The hearing was to resume October 18, but days before it convened Puritan voluntarily agreed to abandon its operation on Ray's property, requested release of the bond it had posted, and asked the state to transfer acreage fees on the remaining portion of its Island Creek permit to new operations in Johnson County.[13]

In addition to lending assistance to Jink Ray and his neighbors, AGSLP activists and antipoverty workers also helped organize a two-day tour of strip operations at the end of July in Bell, Breathitt, Floyd, Harlan, Knott, Letcher, Perry, and Pike Counties, concentrating on areas mined since 1966. "We want the people on the tour to see with their own eyes that the 1966 law has not worked in Eastern Kentucky," explained local resident Morris Sheperd. "It has not prevented enormous mud slides, erosion, stream pollution, and the destruction of property." Altogether, fifty people participated in the tour, including officials from the U.S. Department of Interior, the U.S Forest Service, the Federal Water Pollution Control Administration, and the TVA. The federal officials said they felt compelled to join the field trip because it was a manifestation of "grassroots interest" in the conservation practices advocated by reclamation professionals in Washington. During the tour, they got a taste of the tactics employed by strippers to evade the law and good reclamation as well as the methods used by opponents of surface coal mining. At a number of points the motor caravan came upon excavations in access roads and, in one instance, it was blocked by an armed guard. In Clear Creek, AV Tom Bethell made repeated, dangerous climbs over a freshly bulldozed cut in a dirt haul road and managed to carry most of the inspection party to a ridge where a bench stretched around the mountain top. From that vantage point, the tour members also could see a wrecked shovel, dynamited in April.[14]

Strip operators responded to the Pike County showdown and AGSLP tour by utilizing their influence with police, court, and elected officials. Like their counterparts in the lower South, who blamed outside agitators for stirring up African American communities, coal operators of eastern Kentucky sought to blame antipoverty activists for the militant opposition to strip mining, and they were most able and willing to do this in Pike County. Just after the affair on Jink Ray's land, the Pike County sheriff, a representative of the Small Business Administration, and Robert Holcomb—who was the president of the Pikeville Chamber of Commerce as well as the Independent Coal Operators Association (ICOA)—visited SCEF worker Alan McSurely and Pike County AV field director Joe Mulloy to question them about their work. On August 11, McSurely's home was invaded by commonwealth attorney Thomas Ratliff and fifteen armed deputies, who searched the residence for two hours and confiscated all printed or written material. They arrested Alan and, after discovering that his wife Margaret had worked for the Student Nonviolent Coordinating Committee in 1964, they arrested her too. Around midnight Ratliff and the deputies called on Mulloy and put him under arrest. Days later, Ratliff also had SCEF executive director Carl Braden and his wife Anne arrested for attempting to overthrow the government of Pike County. Anne Braden had never set foot in the county and Carl had been there only once, to get the McSurelys and Mulloy out of jail.[15]

Near the end of August, Robert Holcomb publicly admitted that the ICOA had spearheaded the investigation of antipoverty workers, and Thomas

Ratliff, who was a past president of the ICOA and a candidate for lieutenant governor, explained the organization's concern. "From what I have seen of the evidence in this case," Ratliff said, "it is possible that Communist sympathizers may have infiltrated the antipoverty program not only in Pike County, but in other sections of the country as well." The objective of the antipoverty workers, he claimed, was "to stir up dissension and create turmoil among our poor." At the organizational meeting for a surface-mining division of the national ICOA the next month, Ray Johnson claimed that he was driven off Jink Ray's land by "Ned [Governor Edward] Breathitt, outsiders, and Communists," who were determined to destroy strip mining in the area. But the case against the McSurelys, Mulloy, and Bradens greatly misrepresented the intentions and influence of antipoverty workers. "Because the Appalachian Volunteers are working in Pike and other counties with citizens who are standing up in opposition to strip mining, the Eastern Kentucky strip-mine operators seem to have assumed that AVs are responsible for the opposition," the AVs explained in an August 17 press release, but "this is not the case; for us it is clear that the people of Eastern Kentucky would band together against strip mining with or without our help." A report by the Federal Bureau of Investigation noted that Ratliff's prime interest was "ridding Pike County of the antipoverty workers . . . [for] reasons economic and political: (1) he has made a fortune out of the coal industry and still had coal interests; and (2) he is running for Lt. Governor in the Republican ticket and thinks it is a good issue."[16]

With assistance from constitutional rights lawyer William Kunstler, the antipoverty workers convinced the Kentucky Court of Appeals to declare unconstitutional the 1920 state sedition act, the law they were charged with violating. But the coal operators succeeded in undermining their organizing in the coalfields. Pike County officials put heavy pressure on Governor Breathitt to rid the state of the Appalachian Volunteers, and he recommended the OEO cut off their funding, which agency director Sargent Shriver did on August 18. In September, the director of the Kentucky OEO explained that AV funding was cut due to lack of cooperation and open rebellion on the part of the volunteers against community action agencies. While on the campaign trail, gubernatorial candidate Louie Nunn made a promise to "run SCEF and organizations like it out of the state." After Nunn won the election, legislators established a Kentucky Un-American Activities Committee (KUAC) to help him make good on the promise. By autumn of 1968 the AVs were in serious financial straits and both Appalachian Volunteers and SCEF workers were facing KUAC hearings to determine the level of Communist involvement in their respective organizations.[17]

Earlier in the summer of 1968, strip mining opponents also faced another setback when the Kentucky Court of Appeals reversed Judge Chris Cornett's decision in *Martin v. Kentucky Oak Mining Co.*, ruling that mineral owners had the right to strip-mine without surface owner consent and without paying compensation. The Martins argued that parties to broad form deeds could not have intended that the surface could be destroyed in the removal of minerals, since otherwise it would have been pointless to retain surface rights, and they did not intend "mining" in the deeds to refer to contour strip or auger mining, which was unheard of at the time. Based on this reasoning, the Martins requested that mineral owners be stopped from strip or auger mining any area upon which they had permitted the surface owner to make improvements. But the court saw the case as a question of whether or not the parties to broad form deeds "intended that the mineral owner's rights to use the surface in removal of the mineral would be superior to any competing right of the surface owner." The majority decision claimed that only a small portion of the land in Knott County was improved agricultural land at the turn of the century and a great percentage, including the 90-acre tract of which the Martin's property was a part, was hillside land of no productive value. "The argument that no *farmer* reasonably would have intended his *fields* be destroyed by mining operations must be weighed," the justices asserted, "in light of the fact that there were very few farmers and very few fields involved in the mineral deed transactions." Even when the deeds did encompass bottomland, they argued, the rights to this land were included in the deeds because landowners were willing "to take the chance on future destruction . . . to get the immediate money." By this reasoning, if there were any "estoppel" in the case it would be against the Martins, who built their improvements after *Buchanan v. Watson* advised that strip mining could be done under the broad form deed. "It appears to us that if, as we in substance are holding, the mineral owner bought and paid for the right to destroy the surface in a good faith exercise of the right to remove the minerals," the justices concluded, "then there is no basis upon which there could rest an obligation to pay damages for exercising that right."[18]

There were two dissenting opinions in the *Martin* case. One, by Judge Osborne, argued that the case should be dismissed because the parties had failed to show that there existed an actual controversy falling under the purview of the Declaratory Judgement Act. The other, written by Judge Hill, argued that inasmuch as the parties to broad form deeds did not contemplate strip mining it should not be allowed and, if the "rules of construction" are so distorted to authorize the mining method, mineral owners should be answerable in damages to the surface owner. The majority opinion was contrary to the laws and court decisions of sister coal states, Hill noted, and inconsistent with other opinions of the Kentucky court in similar situations. In 1960, the court of appeals had ruled in *Wiser Oil Company v. Conley, Ky.* that the owner of oil and gas rights had no right to use the water-flooding method of recovering oil without the consent of the owner of the surface because "it was the intention of the parties that oil should be produced by drilling in the customary manner that prevailed when the lease was executed." Even the second *Buchanan* decision declared that the owner of the mineral has the paramount right to use of the surface unless that power was exercised "oppressively, arbitrarily, or maliciously," in which case damages were due the surface owner. "I contend that any major destruction of the surface is oppressive," Hill wrote, "but the majority does not, [and] they should reform the rule in Buchanan so as to delete the word and overrule its opinion in *Wiser Oil*."[19]

Save Our Kentucky

Just hours before Republican Louie Nunn replaced Democrat Edward Breathitt as Kentucky's governor in December 1967, Governor Breathitt issued new strip mining regulations. The revised standards were necessary, he said, based on the Reclamation Commission's recent finding "that an imminent peril to the welfare of the citizens of this commonwealth is created when spoil material is stacked on steep slopes, especially during the coming heavy winter rains and snowfall." The commission had adopted two regulations, one prohibiting contour strip mining of any slope steeper than 28 degrees from the horizontal and the other establishing stricter procedures for revegetating strip-mined land, both of which targeted surface miners in eastern Kentucky. Breathitt's order put the steep slope standard in effect immediately, on an emergency basis, leaving the replanting requirements to take effect in thirty days.[20]

Once Nunn was in office, however, operators had little to worry about as far as old or new regulations were concerned. The new governor left the Department of Natural Resources intact and even reappointed Elmore Grim as the director of the reclamation division. But he sandwiched Grim between superiors and subordinates who owed their allegiance to eastern Kentucky Republicans and coal operators. As historian Marc Landy explains, this move meant that the Breathitt administration holdover had little influence in the choice of area supervisors in each of the state's four reclamation districts, and this imperiled any chance for the enforcement of regulations. Area supervisors determined the frequency of inspection and the content of reports reaching the Division of Reclamation's Frankfort office, where punitive action was decided. When they were chosen for their willingness to go easy on strip operators, oversight was a farce. Not surprisingly, during the Nunn administration the reclamation division failed to revoke a single permit or to deny any company the right to apply for additional mining permits.[21]

In the latter part of 1970, Nunn lost his bid for reelection to the lieutenant governor, Democrat Wendell Ford. But the change in chief executives meant little in terms of the enforcement of strip mine regulations. On the campaign trail, Ford made bold pronouncements about applying adequate standards of reclamation, "even if this means the complete abolishment of strip mining in steep grade terrain." A Ford administration would stop strip mining where reclamation was not feasible, he said, and "completely control strip mining where reclamation is possible." But in July the gubernatorial candidate also secretly met with Kentucky coal operators in a Wise, Virginia, hotel room, putting him out of reach of Kentucky's campaign contribution laws, where he made various promises and possibly requested campaign funds. After the election, Ford kept the corrupt northeast area supervisor Carter Combs on the job until the *Courier Journal* let it be known that it was about to expose him. The new governor hastily called a news conference, blamed the former Republican administration for Combs's shoddy performance, and announced that he was to be transferred. Later, Ford was proactive in making other changes to correct some of the deficiencies associated with the regulation of strip mining in Kentucky, ending the fee financing system and providing the funds needed to hire additional inspectors. Yet throughout his administration, enforcement of strip mine regulations was characterized by continued de-emphasis of noncompliance orders and increased reliance on persuasion of operators to improve their mining and reclamation practices, similar to the cooperative relationship established under Governor Nunn.[22]

The willingness to coddle coal operators exhibited by Kentucky's governors between 1968 and 1972 translated into abuse of the land in the strip coalfields of the eastern part of the state. A 1972–73 report on eastern Kentucky strip mining and regulation concluded that revegetation failures and landslides had declined relatively since 1966, but the dramatic rise in the number of acres stripped nullified those gains. The frequency of landslides on a per acre basis had fallen 63 percent since 1964, but the absolute number fluctuated with the contraction and expansion of the industry. Annual land-slide acreage increased from less than 100 acres in the late 1950s to nearly 1,000 acres in 1965, dropped precipitously in 1966 to around 300 acres as the number of permitted acres fell, and climbed again to nearly 1,000 acres in 1972. Another report, focusing on the effects of strip mining on water quality in small streams of eastern Kentucky, showed large increases in chemical pollutants and sediment. Surface operations in the area had increased concentrations of sulfate, calcium, magnesium, aluminum, manganese, iron, and zinc in streams, the authors of the report maintained, while dissolved minerals and increased turbidity of the water were detrimental to benthic (bottom-dwelling) food organisms and fish reproduction, thereby eliminating some species altogether and reducing the populations of others.[23]

The deteriorating environmental conditions caused by stripping as well as the role surface coal mining played in worsening economic conditions spurred on the opposition movement. AGSLP remained one of the most important groups fighting stripping, but other organizations also began to get involved. At its October meeting in 1969, the forty-year-old Kentucky Conservation Council (KCC) passed a resolution urging that strip mining be banned in the eastern part of the state. Stripping had resulted in the destruction of streams, watersheds, natural scenic beauty, and the blocking of natural routes of access, the resolution claimed, and this was despite fifteen years of efforts by legislators and governors to control stripping through laws and regulations. "WHEREAS it is apparent that the practice of surface mining of coal in eastern Kentucky coal fields is generally adverse to the long-range social, economic, and physical well being of the Commonwealth of Kentucky and its people," the organization declared, the KCC petitioned Governor Nunn to sponsor legislation to outlaw contour strip and auger mining of coal in the mountain counties "and to propose similar action by the Federal Government to be effective in the Appalachian coal mining states."[24]

In November, KCC called a meeting in Whitesburg for December 12, to produce a coalition "to be composed of individuals and organizations who share our opinion that ruining the hills of eastern Kentucky for coal which lies within them is uneconomic and untenable." At the meeting those in attendance were asked to endorse the KCC resolution and each individual or group representative announced their position. Harry Caudill strongly endorsed the resolution and urged vigorous action for its implementation. John Franson, the Midwest representative for the National Audubon Society (NAS), and Mike Flynn, writing in for the Cumberland chapter of the Sierra Club, said their groups were undecided. Jim Butler (of the Frankfort Audubon Society), Lewis Howard (of the Kentucky Mountain Coal Company), and Judge George Wooten (Leslie County) personally endorsed a ban on stripping. Dr. Richard B. Drake, from the natural resources committee of the Council of Southern Mountains, announced his organization's "strong tendency to move in [the] direction of stronger laws to regulate strip mining," but he could not say what its position would be on the resolution by the next meeting. Mrs. W. C. Chrisman (of the Kentucky division of the American Association of University Women) said her organization was making a study of conservation, and Mrs. Clifford Herrick Jr. requested information on the economic aspect of strip mining to be used for a study within her organization, the League of Women Voters. Mrs. William W. Ryle, of the Beechmont Garden Club, did not attend but advised by letter that her members were in favor of the resolution. And Tom Ramsey, of the Pike County Citizens Association (PCCA), declared his group's interest in "working on a broad scale against strip mining." Wayne Davis, a University of Kentucky zoologist, then made a motion to delete "East" from the resolution, so that it would apply to both sections of the state, but this was defeated on a five to four vote. The original motion was then passed eleven to two.[25]

After passing the resolution, nominations were opened for the chair and vice chair of an ad hoc committee "to seek legislation to outlaw surface mining." Caudill nominated the absent Mike Flynn for the position of chair, while Loyal Jones nominated Jim Butler for vice chair, and both were elected. The executive committee consisted of Hal Ritchie (KCC president), Harry Caudill, W. R. Holstein (Sierra Club), David Schneider (then head of the Kentucky Izaak Walton League), Mrs. Robert Cullen (Kentucky Garden Clubs), as well as John L. Franson and Jim Butler. After the meeting, however, Flynn declined to chair the committee. "I'm sure you are aware that the Sierra Club is anxious to offend as few legislators as possible at this time in view of the Wild Rivers Bill in the legislative hopper," Flynn explained to Butler in a letter. "Since the informal view of our Executive Committee reflected this concern and the fact that we do not currently have a confirmed position on strip mining," he wrote, "I had to decide as I did." Butler was made the chair of the committee then, and John Franson accepted the position of vice chair, but the Sierra Club's waffling peeved Harry Caudill. "It is inconceivable to me that an organization like the Sierra Club which was organized to preserve mountains for their beauty and majesty can stand idly and silently by while the mountains in the state are destroyed," he wrote to Butler, since "setting aside a small stretch of scenicly beautiful land is a trifling matter indeed compared to the rapidly advancing ruin of a whole mountain range."[26]

During the summer of the next year, the governments of three Kentucky counties also took the initiative to deal with the strip mining problem. Earlier, in 1967, the Harlan Fiscal Court had unanimously adopted a resolution opposing strip mining as practiced, calling for greater restrictions and better enforcement of the control law. In May 1970, the western Kentucky county of Henderson set a new precedent by declaring strip mining a "public nuisance" and banning it. In June, the Knott County Fiscal Court took a similar step, acting on a resolution to ban strip mining offered by magistrates Dan Wicker, a security guard, and Birchel Smith, an unemployed laborer. County judge Sid Williams, an equipment supplier to strip mine operators, broke a two-to-two tie on the resolution, reluctantly voting "for the majority of the people." The three hundred people packed into the courtroom cheered for five minutes afterward. But the state attorney general had already declared that counties could not act where the state had jurisdiction, and Judge Williams said his vote was practically meaningless. With doubts about the legality of the Knott County action, folks in nearby Leslie County took a different approach later that month. Judge George Wooten and others established the Hyden-Leslie County Planning Commission with the intention of using planning and zoning to bring strip mining under control, since the right of a county to plan and zone for proper land use had generally withstood legal tests.[27]

The Council of Southern Mountains began to play a more prominent role in the opposition after a shake-up in leadership at its Lake Junaluska conference in September 1970, including the election of Warren Wright as the executive director. Wright was a self-described political conservative before his land in Letcher County was threatened by a strip mine operation. His father had sold the mineral rights to Consolidation Coal in 1939, but had forced company attorneys to write into the deed that no portion of the surface would be covered by dumping material on it. In 1962, however, the Beth-Elkhorn Coal Company began to prepare the land for stripping and Wright blocked the way of the bulldozer as his wife Mae sat under a tree with a pistol in her lap. The circuit court in Whitesburg ruled that this interference was doing irreparable damage to the company and sent an officer over to allow the miners to auger for coal on the land. Wright took the case to the Kentucky Court of Appeals and the justices studied the matter for eighteen months and finally ruled against him. "That group of learned reprobates entered an outright lie into the language of their Opinion," he recalled, "declaring that as all the conventional underground mining had been finished earlier, my father could have been granting Consol nothing but stripping rights." Even the case record indicated that underground mining was still

Citizens' League to Protect Surface Rights meeting in Letcher County, Kentucky.
(Courtesy of Phil Primack)

going on under the property in 1962, in the same seam being augered. But
"that fight made me socially alive for the first time," Wright remembered.
"That's when I became a citizen of the United States." As the head of CSM, he
continued to express himself passionately on the issue and became an outspoken proponent of stopping stripping by any means necessary.[28]

In the early part of 1971 some of the same people involved in the KCC ad
hoc committee joined together as Save Our Kentucky (SOK). On January 15,
a group of thirty gathered in Berea and adopted a resolution declaring its
sympathy with the earlier coalition effort and establishing the new organization in its place. "SOK's twin areas of concern," the resolution explained, "will
be directed toward surface mining which degrades our environment, with a
focus on eastern Kentucky, and the tax structure which omits the equitable
taxing of the extraction of our natural resources." The organization was to be
"a politically oriented body without exemption from taxes," with the objectives of outlawing strip mining and enacting a mineral severance tax. In attendance at the January meeting were representatives from KCC, CSM, Sierra
Club, the Louisville Audubon Society, the Garden Club of Lexington, the
newly formed Citizen's League to Protect Surface Rights (CLPSR, Letcher
County), AGSLP, PCCA, and the Appalachian Research and Defense Fund
(ARDF).[29]

At SOK's second meeting participants began taking preliminary steps toward acting as a watchdog over the state's Reclamation Commission. They also circulated a rough draft of a severance tax proposal, which would levy a fee of ten cents per ton on coal, and the organization's temporary chair Jack Weller announced the establishment of a policy committee, which included Warren Wright, Citizens League head Joe Begley, and Alice Lloyd College professor William Cohen. In June, SOK hired James Branscome as its permanent director. Branscome had grown up on a 30-acre farm in Snake Creek, Virginia, attended a one-room elementary school as well as Berea College, and made a name for himself in Washington, D.C., as a member of the Appalachian Regional Commission, advocating the abolition of surface mining and nationalization of the coal industry. At the August meeting, Wayne Davis moved that SOK support a ban on all forms of strip mining in the entire state and this passed with two abstentions. James Rosenblum moved that the organization support a 10 percent severance tax of gross value of the coal produced in Kentucky, a motion seconded by Eldon Davidson, and this passed unanimously. Before adjourning, the board also affirmed a position to seek a legislative end to the broad form deed.[30]

While SOK selected its leaders and defined its objectives, Harry Caudill began to act as an intermediary between Harry LaViers Jr., president of the South East Coal Company, and the National Audubon Society, to find "the best approach to be used in serious efforts to stop stripping for coal in Kentucky." South East Coal was a small, unionized deep mining firm in Letcher County, and LaViers was a third-generation coal operator who had recently won more than $7 million in a lawsuit against Consolidation Coal and the UMW, for conspiring to run small coal operators out of business. In late May or early June, LaViers told Caudill that he would anonymously put up a considerable sum of money to fund abolition efforts, and Caudill contacted the NAS because of their practical experience in the conservation field and large Kentucky membership. After several July meetings in Lexington between Caudill, John Franson, NAS executive director Elvis Stahr, and another NAS official, an understanding was reached about what could and would be done with the money. Any gift or gifts would be used for a public educational program, including a series of television spots, speakers, and pamphlets, but they would not be used for "politics." "It is clearly understood by all concerned," Franson explained, "that the National Audubon Society or its representatives will *not* engage in efforts to influence legislation, support political candidates, or other areas which might jeopardize its status as a tax exempt organization." To pick up the political aspects of the campaign LaViers donated $4,000 through Caudill to supplement the budget of SOK, "with an agreement of a similar sum for each month thereafter for at least six months." He also pledged $10,000 a month for the same period to the NAS, "for use in an anti-stripmining campaign in Kentucky and elsewhere in the U.S."[31]

Activists also continued to organize rallies and petition regulatory officials. In July 1971, AGSLP organized a protest meeting at the Knott County courthouse in Hindman. It began with William Cohen singing "Shenandoah" and "This Land is My Land," and included a slide presentation of supposedly reclaimed mountains by Mart Shepard. Speaking before the group, Bessie Smith declared reclamation "a laugh," since the laws were seldom enforced, and said that it was "left up to the people to protect the land, helping others retain their property and protecting the beauty of the land." In the late autumn, ARDF assisted Joe Begley and CLPSR in petitioning the Kentucky Division of Reclamation to revoke the permits of thirty strip coal operators in Perry, Knott, Letcher, Breathitt, Floyd, Knox, Pike, and Harlan Counties for repeated noncompliance with the control law. According to the reclamation division's own records for 1969–71, some of the companies named had fifteen or more violations against them, with one having at least thirty-three. The petition also claimed "that most of the companies which have 15 or more violations have never been fined by the reclamation agency nor have such companies had their permits revoked," as stipulated by the control act. The Tar Heel Coal Company, for instance, had been cited by the division for twenty-two violations, including working off the permit area, letting water and debris cause soil erosion and a landslide, inadequate silt structures, and excessive bench width. None of these resulted in a fine or revocation of permit.[32]

Strip mining opponents gave testimony at a new round of state hearings too. With a legislative session coming up in January, the subcommittee on natural resources conducted a four-day tour of surface coal mining areas in eastern and western Kentucky and followed that with a public hearing from November to December. Reid Love appeared for the League of Kentucky Sportsmen and claimed that a ban was "economically impossible," but he urged the state "to upgrade and enforce" its control laws. Others forcefully called for abolition, or some form of it. Bessie Smith, Madge Ashley, and Dan Gibson spoke for AGSLP, maintaining that strip mining served only the short-term economic needs of the state and expressing exasperation at the apparent futility of using legal means to control the practice. James Branscome pointed out the large numbers of petitions to ban surface mining circulating in eastern Kentucky, but voiced doubts that such methods would be effective in preventing abuses. And while their spokesperson called for "a complete and total end to strip mining," PCCA members paraded one by one before the

Occupation of Knott County strip mine, 1972. (Courtesy of Robert Cooper)

committee carrying placards with the name of a stream harmed by stripping printed on it.[33]

In December, members of AGSLP, CLPSR, CSM, and SOK met at the Cordia School in Lotts Creek to prepare a strategy for the 1972 legislative session. When the General Assembly got underway the next month, two hundred activists traveled to Frankfort to pressure lawmakers. There was a general consensus among the group that only abolition would solve the problems caused by stripping in the mountains—some of the placards carried by demonstrators read "RECLAMATION IS A DAMN SHAME" and "KEEP OUR COUNTRY GREEN AND EMPLOYED: BAN STRIP MINING"—but, ironically, it was a representative from Bowling Green, in western Kentucky, who filed a bill to phase out stripping. Surface coal mining eroded the tax base, took away jobs, ruined landscapes, and tore up roads, Representative Nick Kafliogis told the activists, all of which meant that strip mining condemned eastern Kentucky to perpetual poverty. Other members of the House and Senate as well as the governor spoke to the crowd, but they made no promises and offered little indication of what action they would take. Later, during the session, the phaseout bill failed to make it out of committee.[34]

Back in the eastern coalfields, on a cold and rainy morning at the end of January, local people took matters into their own hands once again. More than twenty women and a few men walked up a company road to the Sigmon Brothers strip operation at Elijah Fork (Knott County) with the intention of occupying the site. The activists hailed from the Floyd County chapter of the Eastern Kentucky Welfare Rights Organization, Mountain People's Rights, AGSLP, and SOK, and they included Bessie Smith, Eula Hall, Sally Maggard, Doris Shepard, James Branscome, and Mary Beth Bingman, a former AV from Wise, Virginia. Only the women went beyond the company gate, however, believing "that if our men would take the same action the coal operators would unleash violence against them." By 7 A.M. they had stopped work at the operation.[35]

Both latent sympathies as well as real divisions between the protestors and strip miners were apparent at the occupied mine site. Most of the all-male workforce who stayed up on the ridge were friendly, and they acknowledged that they would rather do any other work than tear up the land for a living, but jobs were scarce. Other workers kept the press off the mountain and prevented the women from receiving food, water, dry clothes, or sleeping gear from their supporters for the duration of what became a fifteen-hour vigil. At night, a group of about ten men, who were apparently not regular employees of the mine, tore down the women's make-shift tent and threw rocks at them. Down below, at the company gate, Doris Shepard and her two little sisters were verbally harassed by "thugs" well fortified with liquor. The men "were talking about which one of the little girls they were going to rape," Doris remembered later, and this went on until one of their fellow truckers came over to stand with her and the girls. At some point the group of strip mine employees attacked the men gathered at the gate, including a reporter for the Whitesburg *Mountain Eagle* and James Branscome. When they attempted to flee to a hospital, the workers tried to chase them down. Doris and her sister Shirley walked down the one-lane road to block the way of the miners, but they moved out of the way when the miners revved their pickups' engines. Doris then called the state police and after a trooper went up to the strip mine and reported on the violence done to the men, the women voted to come off the mountain. At the gate they found tires slashed and windows broken on two cars, and Branscome's car was overturned. It was an inauspicious end to what would be one of the last significant protests against strip mining in eastern Kentucky.[36]

chapter six

The Dilemma Is
a Classic One

Opposition to Surface Coal
Mining in West Virginia

The campaign to abolish surface coal mining in Kentucky was the most militant and "popular" of state-level movements. Eastern Kentucky residents employed direct action tactics on a scale unseen in other parts of the strip coalfields of Appalachia, and most of the people physically blocking or shooting at bulldozers there were common people. Farmers, miners, and the unemployed all played important roles in the opposition from its inception, and their efforts were the foundation for a more broad-based state campaign to ban stripping. Although its executive committee included wealthy club women and middle-class professionals, Save Our Kentucky grew out of the grassroots efforts of the mostly poor people in mountain counties, and it continued to be responsive to that constituency. By most measures, however, the movement in Kentucky was not very successful. The state never abolished surface coal mining and it was not until the 1980s that it outlawed stripping on broad form deeds without landowner consent.

The closest any state came to banning strip mining was in West Virginia, in 1971, when legislators seriously considered an abolition bill and then passed a measure that declared a two-year moratorium on surface coal mining in nearly half of the state's fifty-five counties. The movement responsible for this moratorium law began in 1967, when antipoverty workers assisted residents of five southern counties in forming the Citizens' Task Force on Surface Mining to help prompt the legislature to pass a new control act. By the next year the opposition was moribund, but in the latter part of 1970, secretary of state and gubernatorial candidate John D. Rockefeller, divinity student Richard Cartwright Austin, and state senator Si Galperin revived it. Rockefeller put up the funding to establish Citizens to Abolish Strip Mining (CASM), headed by Austin, and Galperin led the fight for an abolition bill in the Senate. Like AGSLP, however, CASM did not succeed in its main objective and, the year following passage of the moratorium law, the group disbanded.

Shortly after the 1971 legislative session, *Mountain Life and Work* editor Barnard Aronson penned a letter to the editor of the *Charleston Gazette*, blaming the defeat of the abolition effort in West Virginia on Rockefeller. The state secretary had announced his position on surface coal mining only thirteen days before the legislature convened, Aronson pointed out, which was not nearly enough time to generate general grassroots support. "His leadership of the campaign in the absence of widespread citizens' organization," he argued, "provided the strippers with a ready-made target and a chance to launch an offensive that made Rockefeller not strip mining the principal issue." Aronson also faulted Rockefeller for emphasizing "aesthetic considerations" rather than the costs of stripping for "citizens who have invested their life's work and savings in a home or farm only to have their property ruined and their water made unfit to drink." The state secretary did begin to cite economists and deal with the issue in broader terms, after the "Citizens for a Right to Earn a Living" lambasted him as a millionaire with no feelings for working people. "But the narrow perspective of beauty v. jobs was an albatross around his neck throughout the campaign," Aronson concluded, and Rockefeller typically attacked stripping as an isolated phenomenon.[1]

Some of what Aronson said in his letter was true. Rockefeller did make a rather tardy public announcement of his support for a ban on stripping. But, as this chapter makes clear, grassroots support was strong during the West Virginia campaign and the secretary of state did not persistently present surface coal mining as only a threat to the beauty of the state's landscape. CASM tapped into the already existing but disorganized opposition, inherited its varied critique, and helped give the opposition a larger public voice. Many people expressed their interest in the abolition bill Galperin introduced in January 1971 by establishing CASM chapters or affiliates, organizing rallies, and circulating petitions. A poll in mid-February found Charleston residents to be in favor of a ban three to two, with women slightly more inclined toward abolition than men. In nearby Fayette County, one of the state's leading strip coal producers, respondents to a questionnaire in the *Montgomery Herald* were overwhelmingly in support of either an immediate or gradual end to surface coal mining. But protests, petitions, and widespread sentiment were not sufficient to affect legislators, many of whom were beholden to the coal industry and were always overwhelmingly against a total ban. Even their votes for a partial moratorium meant little when considered against the lack of strip mines and unlikelihood of any new operations being established in the counties where stripping was prohibited.[2]

To fault Rockefeller for the failure to outlaw surface mining overstates his role in the movement and understates the dominance of the coal industry in West Virginia politics. The campaign to stop stripping in the state failed because citizens could not force their legislators to act responsively or hold them accountable for shirking their duties of representation, irrespective of the opposition's leadership. This breakdown in the democratic process is one of the main reasons why many activists in West Virginia and elsewhere began to take their concerns to the national level. By 1971 a regional movement for the abolition of stripping through federal legislation was taking shape as folks from Ohio, Pennsylvania, Kentucky, West Virginia, and Virginia established links with one another and built multi-state organizations. When courts and state legislatures continued to fail them, and nonviolent civil disobedience and violence against property proved to many to be a dead end in terms of a long-term solution, opponents of strip mining started to focus almost exclusively on getting redress through the United States Congress.

Early Regulatory Legislation

West Virginia was the first state in the nation to pass an act controlling strip mining, but this early legislation was as much a product of deep mine operators' fears of increasing competition from strippers as an expression of concern about the environmental and social problems caused by surface mining. Contour stripping had expanded briefly in the state during World War I and the 1920s, yet in 1938 the only recorded surface mine production was in Brooke and Hancock, the two most northern counties, where five strip pits produced 226,000 tons of bituminous coal. With the onset of production for World War II, however, demand for coal was nearly unlimited and underground mine operators became wary of a rapid expansion of unregulated strip mining. Acting at their behest, in late February 1939, Hancock senator Thomas Sweeney and Brooke County delegate L. Reed Clark introduced similarly worded measures to establish controls on the industry.[3]

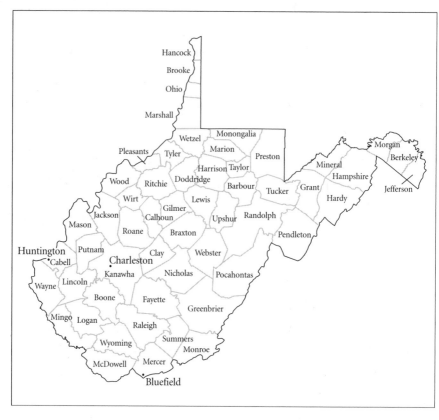

MAP 5. Counties of West Virginia

Senator Sweeney objected to stripping partly on the grounds that it damaged the landscape. "The operators of the strip mines claim that they replace the soil," he explained to his colleagues, "which you and I know not to be a fact." But objections also came from the Brooke County Coal Operators Association, a group representing small deep mine owners, and Sweeney voiced their particular concerns too. "[B]esides disfiguring the land and destroying its fertility," he noted, coal stripping operations stood "in the way of the Coal Operators Association accomplishing what it seeks to do for the miners and the business interest of the county." With the muted but significant support of at least some of the underground segment of the coal industry, the measures to regulate surface coal mining in West Virginia easily passed.[4]

According to the new legislation, controls were necessary because strip mining "causes soil erosion, increases the hazards of floods, causes the pollution of streams of water, causes the accumulation of stagnant waters, destroys the utilization of surface lands for agricultural purposes, creates dangerous hazards in life and property, counteracts efforts for the conservation of soil and preservation of natural resources of the state, and is generally injurious to public health and welfare." To address these many problems, strip operators had to obtain a permit from the chief of the Department of Mines, put up a cash bond of $150 per acre, and replace overburden. "Within a reasonable time to be determined by the department of mines," the law mandated, operators were required "to replace said soil, subsoil or other strata removed from said coal and refill any ditches, trenches or excavations made in stripping said coal." Violations of the act were misdemeanors, punishable by up to a year's imprisonment or a fine of not less than $50 or more than $500.[5]

This first regulatory legislation went into effect in the early part of 1941, but it did not hamper expansion of the strip coal industry. Between 1939 and 1943, production increased tenfold in the state as a whole. It jumped from 50,695 tons to 3 million tons in Harrison County alone, which accounted for nearly half of all strip coal produced in West Virginia during the war. This dramatic growth compounded existing environmental problems and prompted state legislators to strengthen controls on the industry. In 1945 they amended the 1939 act to raise the bond to $500 per acre, with a $1,000 minimum, as well as establish more elaborate postmining regulatory standards, although with some important exceptions. Operators were required to cover the face of the coal "so far as practicable," bury all roof coal and pyritic shales, seal off any breakthrough to underground workings, and provide outlets to conduct storm and seepage waters "with as little erosion as possible." In the cases where stripped lands were previously not used for agricultural or grazing purposes, the director of the agricultural experiment station of West Virginia University could grant an exemption from these provisions. All operators were required to plant trees, shrubs, grasses, or vines, however, within one year after completion of mining and any bonds forfeited for failure to follow the law would be deposited in a fund for reclaiming lands injured by strip mining since the effective date of the act.[6]

Still, the passage of more controls on surface mining put few real constraints on West Virginia's strip coal industry. Provisions for enforcement of the 1945 act were weak, and regulatory agencies faced only minimal protests and isolated public demands to strictly apply the law's provisions. In the late 1950s the mayor of Matewan (Mingo County) complained that surface operations had done "many thousands of dollars worth of damage . . . to the town sewerage system, in addition to losses by private property owners and railroads." This damage was all the more regrettable, he said, because there seemed to be no law that municipalities or private property owners could use to protect themselves. West Virginia conservationists also weighed in on the issue. Strip miners "can go into virgin wood lands and destroy recreational value and scenic beauties," bemoaned the executive director of Sportsmen Unlimited, and this was costing the state millions of dollars every year in lost tourism business. Yet these and other complaints were made outside of an organized campaign or movement, and throughout the 1950s abuses went largely unchecked.[7]

Neither enforcement nor conditions improved much in the next decade and, in 1963, senators and delegates passed more amendments to the state's control legislation that significantly weakened the law. Not only did the new legislation omit the requirement for the operator to indicate the area to be covered by a permit but it also lowered the performance bond to $150 per acre. The consequences of this turnabout were predictable but they were most severe for homeowners living near active operations. When a strip operator began working close to Ellis Bailey's land in 1965 he had all sorts of problems. Bailey woke up one morning to find a bulldozer in the backyard of his Raleigh County home, pushing rocks and dirt toward it. The strippers eventually destroyed his "garden" of four hundred tomato and cabbage plants, ruined his springs and well, and the back of his house fell in from the weight of the spoil washed against it during heavy rains. Not surprisingly, Bailey had little faith in controls. "There's too many poor people being damaged," he said, and "too many lives endangered."[8]

By the mid-1960s, in fact, at least a few West Virginians were calling for the abolition of strip mining. In a letter to the editor of the *Charleston Gazette*, Nicholas County resident F. H. Stewart set out many of the arguments that proponents of a ban would make again in the early 1970s. Confronting the issue of job loss head-on, he noted that most strip mine employees were itinerant workers from out of state, not local. Most of them also possessed the skills for operating heavy equipment, which were easily transferable, so any West Virginian who was put out of work could likely find another job. And closing down surface mines would actually create employment opportunities, when deep mines needing two times as many workers opened to supply coal now being supplied by strip operations. Anyway it was assessed, Stewart argued, there seemed to be no good reason for extracting coal by the stripping method. "Strip mines mar the beauty of our state, destroy vegetation, forest, wildlife in general, erode the soil, pollute the streams, create flood hazards, and, in many cases leave entire mountains of broken stone."

The only gains were made by the "individual spoiler and his well-paid lobbyist, at the expense of all other citizens."[9]

Yet even Ellis Bailey was still willing to give improved regulatory legislation a chance. In January 1967, he and other residents of Fayette, Wyoming, Raleigh, and Boone Counties, along with a few antipoverty workers, formed the Citizens' Task Force on Surface Mining (CTFSM). According to CTFSM's Fayette County chair, Clarendon Williams, the opposition group circulated petitions to most of West Virginia's fifty-five counties calling for corrective measures similar to but even more far-reaching than a new control bill proposed by Governor Smith. First on their agenda was backfilling, followed by prohibition of surface mining on steep mountain slopes and planned stripping, which took into consideration an area's economic conditions, the beauty of the land, and the safety of private property and person. The petitions also demanded a time limit on reclamation, legal assistance to indigent landowners facing damage by strip coal operators, better enforcement through an increased budget for the Department of Natural Resources, and provisions for expanded citizen involvement in oversight.[10]

At the end of January, twenty-five members of CTFSM, including former and current antipoverty workers, took their demands to Charleston, where the Senate committee on natural resources was holding hearings on a new control law. The activists spoke before the committee, presented their petitions, and showed pictures of damage done by strippers in Fayette County. Operators were adamantly opposed to new regulatory legislation, however, and they criticized the bill being considered by the senators, particularly the section allowing the DNR chief to prohibit stripping in "any area which is within 500 feet of any public road, stream, lake, or other public property." O. V. Linde, the executive director of the West Virginia Surface Mining Association (WVSMA) claimed that this provision would eliminate 90 to 95 percent of the surface coal mining in the state. Because of West Virginia's topography, he said, it "would be impossible to operate a surface mine and stay within the limits of this requirement." Other opponents of the control bill declared that the section was an unlawful taking of property and violated the state and federal constitutions.[11]

As finally passed by the legislature, the 1967 regulatory act addressed nearly all of the concerns of CTFSM, including the need for a selective ban on steep-slope stripping and a proposal by Clarendon Williams for triple damages to persons whose land was affected by strip mining. "The legislature finds that there are certain areas in the state which are impossible to reclaim," read the law, "and that if surface mining is conducted in these certain areas such operations may naturally cause environmental degradation as well as imperil life and property." Adopting the wording of Kentucky's regulatory legislation, surface mining was prohibited within 100 feet (rather than 500 feet) of homes, public buildings, schools, churches, cemeteries, commercial or institutional buildings, public roads, streams, and public property. The act also eliminated a requirement that the outer slope of the fill bench of strip mine operations be no greater than a 45-degree angle, requiring instead that backfilling be done "in such a manner as to prevent water from flowing over the outer slope of the disturbed area." Enforcement was given to a Division of Reclamation, within the Department of Natural Resources, as well as a Reclamation Commission, composed of the DNR director, reclamation division chief, and the director of mines. The commission had the task of promulgating "reasonable rules and regulations" and conducting hearings. The law also set up a reclamation board of review, appointed by the governor and consisting of a coal industry representative, a forestry expert, an agriculture expert, an experienced engineer, and a person familiar with water conservation.[12]

Despite provisions making West Virginia's control law one of the toughest in the Appalachian region, regulatory agencies once again faltered in their enforcement duties, and the amount of unreclaimed land actually increased in the following years. By 1971 the state's strippers had created at least 6,563 linear miles of highwalls, benches, and banks. Altogether they had disturbed 250,000 acres of land, three-quarters of which had less than 75 percent vegetation cover, and 109,613 acres of which was classified as not reclaimed for having less than 50 percent vegetative cover. Lands above, below, and across from strip mines had been affected too, on the scale of 3 to 4 acres for every acre stripped. A good portion of the acreage permitted under the 1967 control law was in the southern part of the state—particularly Boone, Fayette, Kanawha, McDowell, Preston, and Raleigh Counties—and this section continued to experience the most severe disturbances. But opposition generated by the failure of the regulatory legislation was not restricted to southern counties. During the late 1960s, an increasing number of residents of all parts of West Virginia became outspoken opponents of surface coal mining, and even outsiders voiced criticism. Violence was boiling below the surface, warned Supreme Court justice William Douglas, and now was the time for the state's legislature to completely abolish strip mining.[13]

Reorganizing the Opposition

In the off-year election of November 1970, Charleston music store owner Si Galperin won a seat in the state Senate largely for his advocacy of a ban on strip mining. In mid-December he attended the annual meeting of the WVSMA, along with Secretary of State John D. Rockefeller, to debate the organization's director, O. V. Linde. In his remarks, Galperin cited the continued problems caused by surface coal mining since 1967, including slope instability and erosion, the consequent threat posed to life and property, sedimentation of streams and increased flooding, and increased unemployment when less efficient deep mines were forced to close or never opened. "[Y]ou would be amazed at the number of people that have written to me, called me, come to see me," he told the coal operators, "to tell me of their personal experiences from these abuses." Some of his constituents were afraid to go to sleep at night "for fear that their houses will not be there in the morning." But Linde dismissed these fears and made an appeal for strip mining under state controls. The "results of our efforts since 1967 convince us that we can strike a balance between use and conservation of natural resources," he said, "and that we can strike that balance under the current legal and regulatory structure."[14]

Following the Galperin-Linde debate, Rockefeller decided to make the fight against stripping central to his campaign for governor and he called a meeting with the senator and his legislative aide, Richard Cartwright Austin. The issue of strip mining was not new to Rockefeller. He had come to West Virginia as an antipoverty worker and, as a state legislator, played an instrumental role in passage of the 1967 control act. Austin also had some experience in the struggle against surface coal mining. He had moved from his native Cleveland to serve as minister to a parish in Clearfield, Pennsylvania, where he could see three strip mines from his home. In 1966, Austin went to West Virginia and his first ministerial post there was in Boone County, where he saw more stripping and witnessed people's opposition to it. While on leave in 1970, Austin decided to join the staff of representative Ken Hechler, the leading opponent of surface coal mining in Congress, and in November he went to work for Senator Galperin. At the late-December meeting called by Rockefeller, Austin agreed to head Citizens to Abolish Strip Mining and the wealthy state secretary wrote out a personal check for $15,000 to sustain the group.[15]

Shortly after establishing CASM, Rockefeller held a press conference to publicly declare his support for a ban. The dilemma was a classic one, he said, between job opportunities and natural beauty, both of which state policy should preserve. Rockefeller suggested that the way to preserve both was by passing a prohibition bill, which he would have Senator Galperin introduce when the state legislature opened on January 13. "I am concerned about jobs," he stated, "but I might note that one reason stripping is so profitable is that it has a low employment factor relative to deep mining." Immediate impact on jobs would be negligible, anyhow, the state secretary claimed, since the industry provided employment to only 3,650 residents, or one-half of one percent of the state's workforce, and those workers possessed easily transferable skills. As to loss of tax revenue caused by a ban, he suggested that a severance tax on deep-mined coal could make up the difference.[16]

L. J. Pnakovich, the president of United Mine Workers District 31, disagreed with Rockefeller, rejecting his estimates of small job losses and questioning prohibition in terms of property rights. The number of workers directly employed by strip coal operations was much higher than 3,650, Pnakovich insisted, and he "doubted if the state could tell owners of mineral rights that they could not mine their coal." The union leader did not deny that surface mining caused environmental damage but, like O. V. Linde, he maintained that the 1967 control act was sufficient to check this if enforced. Miles Stanley, the president of the West Virginia Labor Federation (AFL-CIO) also disagreed with Rockefeller but admitted that strip mining was not sufficiently regulated. "Reclamation," he said, "leaves a lot to be desired." If operators were unable to restore the land to a condition as good as it was when they started, "they should be prohibited from mining." But jobs were involved in abolishing stripping and coal was urgently needed, Stanley argued, while the possibilities for improving regulatory legislation had not been exhausted. In mid-February, the federation's executive board adopted a similar position. With a 6.6 percent unemployment rate, the board declared, absorbing the thousands of strip mine workers made jobless by a ban was wishful thinking. But if new safeguards were not adopted by the legislature and the problem is not solved, the labor body maintained, "the AFL-CIO will join in the effort to ban strip mining."[17]

Other West Virginians agreed wholeheartedly with Rockefeller, and a visible, organized movement to enact state legislation to abolish surface coal mining quickly took shape after his press conference. Raleigh County residents resurrected CTFSM at a rally on January 9, which included Si Galperin as the main speaker. The group's new chairman, Ellis Bailey, claimed that the state's hills were being destroyed faster than ever and the only answer was to stop stripping once and for all. In mid-January Richard Austin opened a

CASM office near the statehouse in Charleston, by which point the organization was already sponsored by eight different groups throughout the state, including CTFSM, the Elk River Basin Protective League, Citizens for Environmental Protection, Cranberry Backcountry Preservation Society, Mountaineers Against Strip Mining, the Montgomery Citizens Antipollution Group, Cabin Creek Citizens Union, and the Public Affairs Conference. Other groups that backed regulatory legislation but not a ban, such as the West Virginia Highlands Conservancy and the Izaak Walton League, also coordinated their activities with CASM.[18]

Through the end of January and into early February, various individuals and organizations made public statements in support of abolition, revealing the many different ways stripping impacted people and the land as well as the numerous reasons to outlaw it. West Virginia League for Better Housing president Carolyn Tillman said her group's support was based on surface mining's adverse effects on housing in the southwestern part of the state. Concerned West Virginians adopted a policy in support of abolition because state officials responsible for enforcement of regulations were unable to "stand up to the surface mining industry and its suppliers and financial and political backers." Some coal miners rallied to the cause of abolition too. In early February a group of twenty Boone County deep miners visited Charleston to lobby their representatives in support of a ban, claiming to represent a wide segment of the workers in their industry. Deep miners were opposed to the position taken by the UMW leadership, they argued, and they recognized that unemployment caused by abolition would be addressed by opening up new underground mines. Strip mining polluted the water, eliminated game animals, and left ugly and dangerous highwalls, said union member Jerry Hughes, and as a rule strip mine employees did not even live in the areas they were destroying.[19]

Boone County was, in fact, a center of rank-and-file miners' activism on a whole host of issues, and CASM concentrated much of its organizing there. On February 4, the abolition group held a meeting in Madison that drew two hundred participants, many of them miners. The highlight of the evening was a panel discussion that included Arnold Miller, a local deep miner and president of the West Virginia Black Lung Association. Responding to reluctance expressed by some members of the audience to holding a massive demonstration in Charleston, Miller supported the idea and rallied underground miners to demand a ban on stripping. Another member of the panel was Ivan White, a member of the House of Delegates from Boone, a coal miner, and black lung activist. He made an emotional appeal for outlawing strip mining, rhetorically asking how a mountain could be torn down and "built back up like God made it." A little more than a week later, some of the same people who attended the Madison meeting gathered in the capitol's rotunda to support legislation amending the state's workers' compensation law. When union leaders walked into the rotunda, rank-and-file miners angrily confronted them about their position on legislation to abolish strip mining. Pnakovich claimed that the union's strip mine membership had to be protected, but the miners pointed out the prevalence of out-of-state workers on strip sites and the damage done to homes and roads by slides.[20]

In addition to the contributions of various citizens' organizations and rank-and-file miners to the campaign for an abolition bill, local and county governments also got involved. In early February, after being addressed by Robert Handley of the Coal River Improvement Association, a ten-to-two majority of the St. Albans city council passed a resolution supporting a state ban. St. Albans drew its water from the Coal River, which was being filled by silt from strip mines in nearby Boone County, and Handley explained that dredging the river would not only be expensive but also only a temporary solution.[21] State officials responsible for conserving West Virginia's soils were also generally in favor of a ban. Replying to a poll of soil conservation district supervisors for the five-county Potomac Valley district, W. C. Taylor said his office's position was that stripping should be stopped. "We believe that the injury and cost in resources and in the population of West Virginia," he wrote, "far outweigh the benefits to the few who desire profit and employment from the operations." A Mingo County soil conservation district supervisor related his difficulty with revegetating stripped lands and questioned the purported benefits of surface mining. Replanting of both grasses and seedlings was limited by the absence of any soil, he said, because "what soil was there is in streams and river beds, so why sell our state for a few dollars?"[22]

As the abolition movement began to gather its forces, however, opponents of a legislative ban on strip mining organized a formidable movement of their own. At a January meeting in East Bank, seven hundred wives of strip mine employees established the Surface Miners Auxiliary of West Virginia (SMAWV), under the leadership of a Mrs. William Strange. "If the abolition bill goes through," Mrs. Strange explained, "it's going to knock thousands of people out of work." On January 20, the SMAWV sponsored a rally in Charleston that drew several thousand opponents of a ban, making it one of the largest protests in the capital's history. The dominant theme of the speeches and placards at the rally was job protection. In league with the SMAWV and

UMW, the coal industry also stepped up its pressure on mountain state residents and their legislators. On January 26, operators released a study which claimed that West Virginia's economy would lose $232 million and at least 23,500 people would be put out of work, counting those directly and indirectly dependent on the industry for employment, if strip mining were no longer permitted. Two days later, UMW International vice president George Titler suggested that Rockefeller's family earned some of its great wealth from Indiana strip mines and declared his opposition to a ban. "I believe anyone who owns a piece of coal land," he said, "has a right to mine it if the land is properly reclaimed."[23]

In early February WVSMA started running television commercials on eight different stations in the state. The spots touted the association's support for strong and enforceable reclamation laws, the ability of its members to enforce standards by self-policing, and its funding of reclamation research. But they also made an economic argument against abolition. The centerpiece of each of the commercials was "average" folks explaining why they opposed a ban, including a welder, two truck drivers, restaurant and service station owners, a grocery store clerk, an auger operator, and Mrs. William Strange. All of these people emphasized the detrimental economic impact that was sure to follow a prohibition on strip mining. In one commercial, heavy equipment operator Jack Burdette expressed his anger at the abolition campaign for threatening to take away jobs without good reason. "Heck," he said, "you can't raise a conversation in these hills let alone a crop."[24]

The "Fair and Equitable" Compromise

Both legislators and activists believed that strip mining would dominate winter proceedings of the House of Delegates and Senate, and a flurry of legislative activity soon confirmed this expectation. At the end of January, Galperin introduced an abolition bill in the Senate, disallowing new permits after June 30, 1971, and prohibiting surface coal mining altogether by 1973. The bill was immediately referred to the Natural Resources Committee, and committee chair Carl E. Gainer appointed a five-member subcommittee, which included Galperin, to study the effects of a ban. The subcommittee initially planned to visit strip mines, to gauge the impact of the 1967 control act, but snow forced members to remain in Charleston. Instead, they heard arguments from representatives of proponents and opponents of abolition in a long closed-door session. Rockefeller and wildlife management professor Robert L. Smith spoke in favor of Galperin's bill, while the WVSMA president and director, Gil Frederick and O. V. Linde, spoke against it. By the middle of February the weather had improved enough for a day-long tour of the steep slopes of West Virginia's southern counties, where the subcommittee was confronted by sign-toting advocates of outlawing strip mining.

In addition to Galperin's bill, two senators introduced a proposal giving state residents the opportunity to make complaints against reclamation supervisors or inspectors who willfully or deliberately did not enforce regulations, and allowing courts to order officials to rightfully perform their duties or be tried as criminals. Another legislator brought a bill to the Senate floor that allowed West Virginians to vote to outlaw stripping on a county-by-county basis, with a five-year time lag after such a vote. Members of the House also proposed separate measures strengthening regulations and outlawing stripping. A Monongalia County delegate introduced legislation extensively revising the 1967 act, providing for public hearings prior to issuance of permits, requiring the maintenance of uniform records, reconstituting the reclamation commission, and increasing bonds. On the same day, Berkeley County delegate Robert Steptoe and his Wyoming County colleague Warren McGraw introduced an abolition bill, similar to Galperin's proposal in the Senate.

With various proposals up for consideration by the legislature, on February 15 a crowd of one thousand, consisting largely of abolition supporters, gathered at the statehouse. Tensions ran high that evening and state troopers broke up a number of scuffles. At one point strip miners yanked an "Abolition Only" sign from the neck of a young women, and others hurled a full soft drink can at the head of a *Charleston Gazette* photographer. Outside of the capitol building Rockefeller led a parade of sixteen conservationist and citizens' groups, denouncing strip miners for bringing "short term benefit but long-range detriment to all of us living here." Representatives of the West Virginia Highlands Conservancy, Izaak Walton League, and Concerned Citizens for West Virginia, all of which had eventually decided to support abolition, echoed the state secretary's sentiments. And rural mountain people like Martha Sexton of Cabin Creek emphasized the damage done to property and livelihoods by surface coal mining. "It took bulldozers thirteen days to get the slide out [from her home]," she said, "and it destroyed everything."[25]

Two days after the February 15 gathering, the Senate natural resources subcommittee voted four to one against recommending Galperin's abolition bill, and they canceled a planned tour of strip mines in northern counties in order to give full consideration to a control bill. Chairman Pat Fanning from McDowell County said the majority of the subcommittee took the view that strip miners were not causing "unacceptable waste," as stated in the abolition legislation, and stripped lands could be converted for use as "airports, subdivisions and other things." When Senate natural resources chair Carl Gainer reappointed the subcommittee to draft new legislation he ensured the dominance of this view by replacing Galperin with Senator Chester Hubbard of Ohio County.[26]

Despite the quick defeat of the abolition bill in committee, however, proponents of a ban continued to organize meetings and rallies, including one in Morgantown that drew a thousand demonstrators, and these events continued to shape the thinking and work of legislators. Another part of the backdrop for the deliberations at the statehouse was a developing controversy over the DNR's granting of a stripping permit near the Kanawha Run campground on Sutton Lake, in Braxton County. In mid-February, Colonel Maurice D. Roush, chief of the Huntington district of the U.S. Army Corps of Engineers, had raised questions about the permit, noting the heavy load of silt that the lake had received during the past summer. Robert Flint, head of a seven-county tourism group called Heartland, also objected to the stripping because "it is taking place in one of the most scenic areas of the county." But after DNR land reclamation chief Benjamin Greene toured the 10-acre operation with Major John Hill, also of the Huntington Corps of Engineers, they announced that the mining would go forward and issued plans to minimize any adverse impacts on Sutton Lake.[27]

On February 21, as part of a weekend of rallies throughout the state, CASM organized a demonstration at the Kanawha campground that drew two hundred people. Richard Austin and Sutton Mayor O. L. Holcomb addressed the crowd, as did John D. Rockefeller. The next day, abolitionists brought to Greene petitions against the Sutton permit, bearing the signatures of one thousand Braxton County residents, as well as resolutions from the towns of Sutton and Gassaway, which also opposed the strip mining. While in Greene's office, Robert Flint said he listened in on a telephone conversation between an aide to Governor Moore and the reclamation chief that revealed the insincerity of state leaders. Supposedly, the aide advised playing the situation by ear, telling the protestors "anything to get them off our backs," and getting "the boys" to pull out of the job "until this group, the colonel [Roush] and press die down." Neither the Moore staff member nor Greene knew Flint was on the phone and they disputed his recounting of the conversation, but on February 24 the permitted strip mine company voluntarily suspended its work pending a two- to three-week review by the DNR.[28]

As Braxton residents were pressing their particular case, Richard Austin and other CASM members continued to address legislators on a statewide ban of strip mining. On February 25, Austin demanded a roll call on abolition bills in the full Senate and House and he put his organization on record as favoring federal legislation to abolish stripping nationwide. But the bill recommended two days later by the Senate natural resources subcommittee was not an abolition bill. It set a thirty-foot limit on highwalls and more stringent bench width measurements, required that spoil be retained on the bench or only according to a specified plan, and increased the permit fee and raised the performance bond. Activists responded to the proposal with a protest, barring entrance to the Senate chamber. The crowd included residents from Kanawha, Boone, McDowell, Mingo, and other counties, and they joined together in a rousing rendition of "The West Virginia Hills." One participant in the civil disobedience was retired deep miner James Washington, who rejected the notion that stripped land could be reclaimed as well as the argument that abolition would cause job losses. Strip mining was a "fast way to make a buck," he said, and every strip mine employee displaced three or four deep miners.[29]

Back at the DNR offices a storm was still brewing, partly due to the Sutton Lake permit and partly due to the outspoken acting deputy director Norman Williams. In February, Williams had become the first member of the Moore administration to call for a ban. "[I]t seems as if the state is allowing the operator to use his property in such a way to deprive his neighbors of the use and enjoyment of their property," he said in a speech to the Kanawha Valley Unitarian Fellowship, "and this, at least to the victims, seems like an unlawful taking of property." In March, Williams announced that six of the agency's ten division chiefs quietly supported outlawing surface coal mining and he called for the resignation of committee chair Carl Gainer. Williams accused the senator, who was also an oil company distributor, of a pro-industry bias. "I believe there exists no greater obstacle to the enactment of legislation which will truly protect the environment of West Virginia," he said in a speech delivered on the Senate side of the capitol rotunda, "than the leadership of the senate natural resources committee." Other members besides Gainer also had links to extractive industries, including Senator Tracy Hilton, the largest strip miner in the state.[30]

In the legislature, Galperin tried once again to bring abolition before his fellow lawmakers in the form of an amendment that would have phased out the industry in two years. Senators from Fayette, Monongalia, Cabell, Logan, and Mingo Counties voted for the Galperin proposal, but it was rejected on a twenty-seven to six vote. More than thirty other amendments met a similar fate. A measure introduced by Senate majority leader and Kanawha Democrat W. T. Brotherton, however, was defeated only on a tie vote of seventeen to seventeen. The next day, on a motion by Senator William Sharpe, the Senate reversed itself and adopted the Brotherton amendment to the control bill on a vote of twenty-four to nine. This change permitted stripping where the heaviest mining was already taking place (Kanawha, Raleigh, Fayette, Boone, McDowell, Logan, Wyoming, Mingo, Lewis, Preston, Barbour, Greenbrier, Nicholas, Harrison, Mineral, Upshur, Grant, Randolph, and Monongalia Counties), allowed limited stripping with no new permits for one year in others (Clay, Mercer, Webster, Summers, Brook, Tucker, Wayne, Taylor, Putnam, Marion, Gilmer, Braxton, Cabell, and Hancock Counties), and banned surface mining for a year in the remaining counties, areas which were classified as unsuitable for mining or where there were no minable reserves. During the year-long moratorium experts would study "the damaging and economic effects of strip mining with a view toward deciding what to do about the industry in the future." Senator Sharpe said he had made the motion because he was opposed to abolition and believed a thorough study would show that surface coal mining could be properly controlled.[31]

On March 6, however, the House Judiciary Committee recommended a modified version of the Senate bill, with the Brotherton amendment removed. Days later, the full House engaged in its longest debate on record (at just under five hours) before sustaining the committee recommendation and rejecting a partial moratorium on a vote of fifty-five to forty-four. The Brotherton amendment would "deprive several thousand families of their livelihoods just overnight," said judiciary vice chairman and Raleigh delegate Anthony Sparacino, and it was "nothing more nor less than an effort to condition the surface mine operators to total abolition." But in the wee hours of the morning, the House passed a bill proposed by Fayette County delegate George Seibert that permanently banned strip mining from counties where it was not yet being done. After a conference committee worked out a compromise measure, the West Virginia legislature passed a two-year halt on all strip mining in the state's twenty-two unstripped counties (half of which had no minable reserves), with the Senate voting twenty-four to nine and the house of delegates voting eighty-eight to four. The final bill also increased performance bonds to $560 per acre, required the construction of approved drainage systems before any operations could begin, and mandated delayed blasting techniques. Perhaps indicative of the weakness of the legislation, the president of the wvsma praised the legislation as "fair and equitable," and while signing the bill Governor Moore suggested that, considering "the times and temper," the new control act was a good one.[32]

Development of a Regional Movement

When Senator Brotherton first proposed a halt on strip mining in selected counties he meant for the ban to be temporary, to allow for a study of the practice. The bill that came out of the conference committee in March established a two-year partial ban and Senate Resolution no. 37 directed the Government and Finance Committee to make an investigation, which it did by contracting with the Stanford Research Institute (SRI). But CASM questioned the choice of SRI for the study, suggesting that the research institute could hardly be expected to issue a report that was not biased against prohibition. The president of SRI, Charles A. Anderson, was on the board of directors of Continental Oil, which owned the Consolidation Coal Company. In 1970 Consolidation was the second largest producer of strip-mined coal in the nation, including 1.3 million tons in West Virginia. The chairman of the institute's board, Ernest C. Arbuckle Jr., was on the board of Utah Construction and Mining Company, the fourth largest producer of stripped coal in the country. And many of the institute's other board members also had ties to energy producers, including the chairman and chief executive officer of the Montana Power Company, a director of the Shell Oil Company, the chairman of Tenneco, Inc., two directors of the Southern California Edison Company, and a director of Union Oil. Still others had a stake in the manufacturing of equipment used at strip mines or in railroads that transported surface-mined coal.[33]

As expected, the SRI report on surface coal mining in West Virginia was biased, but the bias was hidden by judicious language. In an assessment of past legislation SRI acknowledged the failure of the 1939 act to control strip mining abuses and described the 1963 regulatory legislation as "a major step backward in terms of control over surface mining and reclamation." Passage of the 1967 control act was followed by compliance with the law's new requirements, the report asserted, and the treble damages provision provided "an effective and immediate monitoring remedy for direct damages," such as from flyrock, although it was not effective in dealing with indirect damages, such as those caused by flooding. This limited but proven efficacy of regulation, along with certain aspects of the deep mining industry, ruled out either a total ban or phaseout of stripping statewide. It would be impossible to place all strip mine employees in new jobs in the event of an immediate phaseout of strip mining on steep slopes, the report explained, and even a gradually instituted ban on steep slopes would cause severe economic problems in southern West Virginia, particularly in the eight counties where more than 60 percent of all surface coal mining was concentrated. In addition, it would be difficult for deep mines to expand their production to compensate for curtailed stripping. The deep mine coal industry faced a shortage of skilled labor, rising wages, and decreased output as the result of new health and safety legislation, current technology, and controlled prices.[34]

The conclusions of the SRI study lent support to legislators' rejection of a ban on strip mining and helped undermine the sense of purpose of the abolition movement, which was already being fractured by internal dissension. With the 1971 legislative session finished, Richard Austin closed CASM's Charleston office, though the organization was still around the next year, publishing its monthly newspaper and pushing for a statewide ban on strip mining. During the 1972 legislative session, Senator Galperin proposed another abolition bill, one that discontinued the issuance of permits to mine new areas after June 1972, as well as a companion bill calling for special job placement for displaced strip mine workers. CASM affiliates mobilized in support of these measures, but their organizing was not comparable in scale or intensity to what they had done the year before. Proponents of outlawing surface mining wrote and visited their representatives but they had only one rally of any note. Subsequently, Galperin's proposals died in committee and CASM dissolved for good.[35]

Yet strip mining was a salient issue during the primary and general elections in 1972. In the May 9 primaries, Rockefeller trounced two other Democratic candidates for governor, both of whom were heavily backed by coal industry money. In the primary race for the fourth congressional district, West Virginia residents chose prominent abolitionist Ken Hechler over James Kee, one of the coal industry's most reliable supporters, who was also backed by the UMW. Strip mining opponent and delegate Warren McGraw challenged and defeated Tracy Hilton in a state senate race for Raleigh and Wyoming Counties. Boone County delegate and stripping foe Ivan White fended off a challenge from a strip mine equipment dealer, spending only $1,000 to the dealer's $100,000, and despite much evidence of voting fraud. In Kanawha County alone, which then included 15 percent of the state's population, eleven abolitionists emerged as nominees for the county's fourteen seats in the House of Delegates, including eight Democrats and three Republicans. Across the state, a dozen more CASM-endorsed candidates won primaries for House seats, and three abolitionist candidates for the Senate, not including McGraw, won their primary races. Most of these advocates of prohibition went on to victory in November, although Rockefeller was soundly defeated by Moore. Believing this loss to have been the result of his abolitionist position, Rockefeller eventually became a strong defender of the strip mine industry.[36]

Grassroots organizing against surface coal mining continued as well. Even as CASM dissolved, some activists made connections with abolition advocates in other states as a larger, regional movement started to evolve. The West Virginia group had sent a representative to a Save Our Kentucky meeting in the spring of 1971, and later that summer Richard Austin facilitated a gathering in Ohio. The Ohio meeting was sponsored by Concerned Citizens Against Strip Mining and included participants from Pennsylvania as well as the Sierra Club's eastern representative, Peter Borelli. In the fall, representatives from SOK, CASM, Stop Ohio Stripping, and the Wise County (Virginia) Environmental Council met in Huntington, West Virginia, and formed the Appalachian Coalition to coordinate the regional movement for a ban. Despite the failure to outlaw strip mining in West Virginia and Kentucky, then, and the toll this failure took on opposition groups, the abolition movement did not collapse in the early 1970s. It regained strength with the development of an organizational structure for a regional movement to enact a national ban.[37]

chapter seven

Liberty in a Wasteland
Is Meaningless

Strip Mining Opposition at
the Federal Level, 1968–1972

T he development of a movement demanding federal action on surface coal mining in the 1960s and 1970s occurred against the backdrop of change in the stripping
industry, including new patterns of ownership and increased levels of production. Since World War II, bigger
coal companies had been engulfing smaller companies to reduce competition and generate larger cash flows. In the decade after 1955, Peabody, Consolidation Coal, Island Creek, and Pittston acquired thirty-one smaller firms
among them, giving each important stripping operations in Ohio, Kentucky,
and West Virginia. Such mergers made the larger coal companies prime tar-
gets for takeovers themselves, and by the mid-1970s many had been bought
up by energy conglomerates. With the depletion of energy resources in the
mid-1960s, manufacturers, utilities, metal producers, and especially oil producers used their excess capital to invest in the capital-starved coal industry.
Continental Oil picked up Consolidation Coal, Kennecott Copper acquired
Peabody, and others like Occidental Petroleum and Standard Oil also made
coal a target for diversification. At the same time, eastern utilities attempted
to free themselves from reliance on independent producers charging spot-
market prices by signing long-term contracts with the large energy corporations and expanding their reserves. By 1976, TVA alone controlled 412 million
tons of coal reserves.[1]

The multiple mergers of the postwar period combined with the rise of coal prices to stimulate expansion of the strip mining industry and increase production, especially during the so-called energy crisis in the early 1970s. Surface mines produced 40 percent of the nation's coal by 1970 and surpassed underground mines in total tonnage three years later. Although anthracite stripping continued to decline as accessible reserves were depleted, bituminous and lignite surface production rose by more than 30 percent between 1970 and 1976, from 260 million tons to 380 million tons. In the first part of the decade, Kentucky led all other states in production, followed by Pennsylvania, Ohio, Illinois, Indiana, and West Virginia. This ranking was altered slightly by 1976, when Wyoming was the third-leading producer, a change that was indicative of the movement of surface coal mining westward. Less important in terms of total tonnage were such eastern states as Virginia and Tennessee, which produced 13.9 million and 4.8 million tons of strip coal, respectively, at mid-decade. But surface mining was concentrated in only a handful of southwestern counties in Virginia and eastern counties in Tennessee, and its impact on these areas was as significant as in the states where production was much greater (see Table 1).[2]

The changes in the strip mining industry after mid-century exacerbated the problems caused by states' failures to effectively control or outlaw stripping, and they made federal regulatory and abolition legislation appear even more necessary to activists and lawmakers alike. Everett Dirksen had introduced the first federal control bill in 1940, and several members of Congress introduced bills in the 1950s and 1960s to institute surveys of the damages done by strip mining, but the first full-scale congressional hearings on the issue were not held until 1968. Between 1968 and 1977, when Congress finally passed the Surface Mining Control and Reclamation Act (SMCRA), there were more hearings and numerous legislative proposals. The bills introduced by members of Congress in those years ranged from weak federalist arrangements, in which the states would retain much of their autonomy and oversight authority, to a phased-out ban on surface mining. Debate on control measures centered around issues like the extent of state sovereignty, enforcement authority and criteria, surface owner rights, standards for reclamation, and provisions for an abandoned mine reclamation fund. In the oftentimes heated exchanges during hearings, witnesses and committee members also discussed the need to balance energy demands with environmental concerns.

Congressional action in the late 1960s and 1970s was largely a response to increasing public pressure. By the early part of the 1970s, a number of state governments had refused to ban strip mining and courts had issued judg-

TABLE 1. Top Ten States in Bituminous and Lignite Surface Mine Production,
1973 and 1976

	1973		1976
State	Quantity (1,000 Short Tons)	State	Quantity (1,000 Short Tons)
Eastern Kentucky	33,413	Eastern Kentucky	50,587
Western Kentucky	31,337	Western Kentucky	28,913
Pennsylvania	30,195	Pennsylvania	42,018
Ohio	29,558	Wyoming	30,312
Illinois	29,002	Ohio	29,956
Indiana	24,465	Illinois	27,231
West Virginia	19,932	Indiana	24,931
Wyoming	14,461	West Virginia	21,275
Montana	10,724	Alabama	14,131
Virginia	10,524	Texas	14,063
New Mexico	8,330	Virginia	13,940

Source: U.S. Bureau of Mines, *Minerals Yearbook, 1973, Vol. 1: Metals, Minerals, and Fuels,* 336–38, and *Minerals Yearbook, 1976, Vol. 1: Metals, Minerals, and Fuels,* 354–62.

ments that favored strip operators. For the most militant activists, this left either stepped-up nonviolent direct action, violence, or federal abolition legislation as the only means by which to get redress for their grievances. Speaking at House and Senate hearings in 1971, Save Our Kentucky director James Branscome warned subcommittee members that people in eastern Kentucky threatened to use their guns if the political process failed once more. Exaggerating for effect, he claimed that strip miners were making revolutionaries out of mountain people. But the presence of the SOK leader and a number of other abolitionists at the hearings demonstrated their hope, however small, that Congress would finally hear their demands and act on them. Through the early part of the decade, Branscome and his fellow activists organized themselves and lobbied House and Senate members to outlaw strip mining. As this chapter demonstrates, this effort reached its peak of strength in early 1972, when the movement was growing and spreading, national environmental groups were at least rhetorically supportive of a ban, and a significant number of members of Congress signed on as co-sponsors to abolition bills. By the end of that year, however, the campaign began to decline. As the next chapter shows, many activists grew tired and disillusioned as pragmatic national environmental leaders came to the fore and backed away from more radical positions. Members of Congress reacted to both of these developments by focusing more intently on passing only minimal regulatory legislation, which is what they achieved with SMCRA.

From Studies to Hearings

The first federal bill directed at problems associated with surface coal mining was proposed in 1940 by Everett Dirksen, a very conservative Republican senator from Illinois. At the time, his state was by far the leader in national strip mine production and area mining was a significant competing land use with agriculture there, a fact that was reflected in the regulations outlined in the proposed legislation. Dirksen's bill, H.R. 10079, required surface miners to backfill excavations with spoil to return the land to a somewhat level condition. It also required the posting of a performance bond to cover the cost of reclamation, the amount of which was to be determined by the U.S. Department of Interior. Together, these provisions would prevent strippers from leaving behind a rippled landscape, impassable to farm machinery. Yet Dirksen's proposal failed to make it out of committee, and during the next two decades regulatory proponents focused their attention exclusively on passing state legislation.[3]

The second attempt to use the powers of the national government to deal with strip mining was prompted by a 1958 letter from Perry Walper, chairman of the Pennsylvania Conservation Committee of the Allegheny County Sportsmen's League, to House member John Saylor. Walper requested Saylor's participation in bipartisan sponsorship of a federal bill to reclaim abandoned strip mines, although he and his fellow sportsmen would soon be pushing a regulatory bill in the Pennsylvania legislature. Saylor discussed reclamation legislation with staff members of the Department of Interior, but the following year he introduced a bill to provide for a study by the Department of Interior (DOI) instead. Wayne Aspinall, chair of the House Committee on Interior and Insular Affairs, stalled on scheduling meetings for this bill, and both the DOI and Department of Agriculture failed to file reports on the legislation as Saylor requested. The Pennsylvania representative tried to pass another study bill in 1961, however that proposal also never made it beyond committee. Conservationists continued to write Saylor, asking him to initiate some sort of federal response to strip mining, but he had apparently given up on sponsoring even a study bill.[4]

The chances for federal action improved somewhat after 1960, when President John F. Kennedy took office and appointed Stewart Udall as Interior secretary. Unlike his predecessor, Douglas McKay, Udall sometimes seemed as interested in the conservation of the nation's natural resources as he was in their development. Speaking at the first annual Conservation Congress in Kentucky in 1961, he compared strip mining to the soil erosion crisis of the 1930s and suggested the need for federal controls. Addressing the U.S. Congress on the subject of conservation in the spring of 1962, President Kennedy himself announced that the DOI would soon be taking practical initiatives to begin to confront some of the environmental degradation created by surface coal mining. "A serious problem of land conservation calling for immediate attention," he said, "is the serious erosion and river pollution created by surface-mining practices. Techniques must promptly be devised to prevent or minimize this despoilment if we are not to abandon great areas of scenic beauty and create difficult silting problems in many sections of the country." To improve stripping methods and as a first step toward federal control, the president directed Secretary Udall, working with appropriate federal and state agencies, "to recommend a program of research and action." According to a White House spokesperson, this program would start with legislation initiating a DOI study. If the study revealed the need for it, he said, the Kennedy administration would propose other legislation establishing federal regulatory authority over strip mining.[5]

The DOI delayed making a survey of strip mining, however, and Ohio senator Frank Lausche began introducing study bills of his own. As governor of Ohio, Lausche had played a leading role in the passage of the state's first regulatory legislation, in 1947, and he carried his concerns about the problems associated with surface coal mining to Congress. In 1962, he introduced a proposal to authorize a study of stripping operations that would have included their extent, hazards to the public health and safety, and the effect of strip mining on scenic features, fish, and wildlife. An identical bill was also introduced in the House, where it met the approval of both the secretary of interior and the secretary of agriculture. But members of the coal industry appearing before a House subcommittee spoke against the measure, calling the study a needless expenditure of a million dollars and raising the specter of over-regulation. "Industry is properly concerned over the dangers inherent in all efforts at regulation or study by any governmental agency," explained a Kentucky Coal Association representative. "[T]here is always the danger that such efforts, however well-meaning, may result in ill-considered action which could impose economic hardships on industries which, due to the limitations of their number and size, have only a limited ability to defend themselves." With such staunch opposition from coal operators, the bills died in committee. Lausche and House members proposed study bills again the following year, but they met a similar fate.[6]

Viewed by his constituents and residents of other states as a government representative willing to challenge strip miners, Senator Lausche received many letters in the early 1960s asking him to do something to rein in the industry. One Coshocton, Ohio, sportsman inveighed against the destruction of wildlife habitat and pollution of waterways by strippers. "Surely if this desecration comes under no written law," he suggested, "it is a violation of every man's right to the beauty of God's earth." A group of young boys, also from Coshocton, expressed their opposition to Peabody Coal's strip mining in the area with letters and petitions. Brian Winters and John Tozer, ages nine and thirteen, had seen commercials on television advising them against littering to keep America beautiful, but they wondered about the responsibilities of industry. "If we stop being Litter Bugs we can keep America clean," they wrote, "but with this company strip mining it won't be beautiful." Peabody had purchased the land around one of the local public swimming pools with plans to strip it, and John had circulated petitions in response. He sent twenty signatures to Lausche and promised plenty more "if you need it to pass the bill." The executive director of the Izaak Walton League also wrote to Lausche, sending along resolutions passed at the organization's 1964 national convention. The league had come out in favor of passage of the study bill the senator had introduced as well as enactment of adequate legislation to establish controls on strip mining on public lands.[7]

Buoyed by public support, Senator Lausche remained steadfast in his efforts to initiate a federal study and he was successful the third time around, in 1965, when his bill was incorporated into the Appalachian Regional Development Act (ARDA). The ARDA was partly a response to a wide-ranging 1964 report by the President's Appalachian Regional Commission, which had included a call for study of surface coal mining "with a view to identifying appropriate and practical measures to minimize adverse effects of mining operations." The commission worried that a shift from underground mining to contour strip and auger mining would exacerbate unemployment in the region, in addition to degrading the natural environment. As passed by Congress, the ARDA established a permanent Appalachian Regional Commission (ARC) and specified major programs for vocational aid, improvements in public health, highway construction, as well as development of land and timber. It also directed the secretary of interior to study surface mining across Appalachia. Following this directive, Secretary Udall established a study committee from within the DOI and the committee sent questionnaires to mine operators, consulted state officials and conservation organizations, and visited 693 randomly sampled surface mining sites. By 1966, enough of this work was completed to issue an interim report to the ARC.[8]

The interim report claimed that 740,000 acres of land in Appalachia had been directly impacted by strip mining for bituminous coal, in addition to 59,000 acres of strip-mined lands in the anthracite region of Pennsylvania and 74,000 acres disturbed by mine access roads. Most of the surface-mined land was in areas with slopes of less than 17 degrees, but nearly a third of the acreage was on hillsides with slopes between eighteen and 27 degrees, and 90,000 acres had slopes between 27 and 45 degrees. Only 282,920 acres of the nearly 800,000 acres directly impacted by stripping had been completely reclaimed by operators. Mining activity on steep hillsides and the failure of strippers to restore most of the land had caused massive slides more than 600 feet wide on approximately 1,400 miles of bench, as well as more slowly eroding spoil. In addition, over 60 percent of the small headwater streams examined during the on-site surveys had their channel capacities affected by sediment. Of one thousand pH measurements taken on spoil banks, the interim report claimed, more than 80 percent showed a pH of less than five, indicating that they were acidic. Water samples taken 1 to 2 miles downstream from sites showed that more than half of the small streams had pH values of five or less, and a third of the stream channels had deposits of "yellow boy," an iron precipitate that kills aquatic life.[9]

According to the study committee, the large tracts of unreclaimed land resulted "from past failure to recognize reclamation as a necessary part of the cost of mining and of the products resulting therefrom." Through either ignorance or apathy—"public indifference" as the report put it elsewhere—society had accepted erosion, acid drainage, lowered water quality, and other detrimental aftereffects as the costs of "progress." But now, judging by protests against land despoilment, it seemed that the public was willing to pay for reclamation. To remedy the problems left from past operations, the committee proposed federal aid to fund "basic reclamation" of mining sites and haul roads. It also advised Congress to take the necessary steps to protect the public interest if individual states neglected to provide for adequate controls within a reasonable period of time. The study group recognized the diversity of Appalachia and it was hesitant to impose regulatory standards on a regional basis, but it did recommend the designation of a central agency to administer federal activities. The committee also proposed that "mining should be prohibited in areas where reclamation is considered economically unfeasible."[10]

Senator Lausche used the release of the interim report as an opportunity to introduce a bill for extending controls to the forty states that had no strip mining regulations. By 1966, residents of various parts of Appalachia were telling him to push legislation that would provide federal protection of some sort, either through uniform standards or abolition. Donald McIntosh wrote Lausche from Fisty, Kentucky, "begging for federal aid" to stop the destruction of private property, the pollution of streams and wells by acid mine drainage, and the job loss caused by the decline of deep mining. From nearby Hazard, Robert Ritchie sent a letter to the senator calling for abolition to "save our much needed Timber and Bench Land which is our Source of pasture land." Partly in response to such letters, Lausche introduced S. 3882, which directed the secretary of interior to create a national advisory committee to draft federal standards for mining and reclamation and allowed for regulation in the absence of an approved state plan. Hearings on the Lausche proposal, scheduled for August 1967, were postponed at the request of Secretary Udall, yet the DOI staff appeared to be divided. When Assistant Secretary of Interior Cordell Moore spoke at the strip mining symposium in Owensboro, he noted the failure of states to regulate the industry, commented favorably on federal controls, and made mention of S. 3882.[11]

The first important hearings on federal regulation of surface coal mining did not take place until the end of April and beginning of May 1968, when the Senate Committee on Interior and Insular Affairs considered three very similar bills. The administration's bill, S. 3132, was drafted by the DOI and introduced by Senator Henry Jackson for himself and Senators Nelson, Lausche, and Anderson. Senators Nelson and Lausche also introduced bills of their own, providing for the reclamation of previously mined land, but since they had joined Jackson in sponsoring the DOI proposal the whole committee focused its attention on S. 3132. Based on the ARDA study, the administration's bill gave the states "initial and primary responsibility for regulation and control of future surface mining operations within their borders, and for making provisions for restoration and reclamation." It authorized the secretary of interior to appoint an advisory committee that would assist state agencies in developing and administering state plans for the regulation of strip mining and reclamation, and directed the secretary and committee to develop and implement regulations if a state failed to submit its own plan within two years. Once a state submitted a plan acceptable to the Interior secretary, however, federal regulatory intervention would cease.[12]

At the hearings, Secretary Udall defended S. 3132 from its critics on both sides of the controversy over regulation, emphasizing the need for balance in terms of free economic activity and regulatory action. Surface mining had costs that did not always appear in the marketing of coal, he said, costs which "arise with the diminishing of the useful availability of land—with pollution and the hazards to human life, property, and wildlife—with the impairment of natural beauty—with the degradation of other natural values which occurs." On the other hand, the DOI believed that a single set of national standards to address these costs would be "impractical and undesirable." Hence, the bill the agency submitted for the Johnson administration proposed only general criteria as opposed to precise mining and reclamation requirements. Even with this acknowledgment of strip operators' interest, however, the bill drew the opposition of industry representatives, from both the National Coal Policy Conference (NCPC) and the American Mining Congress (AMC). Speaking for the AMC, Joseph Abdnor related the great efforts made by the strip coal industry to reclaim land as part of preliminary remarks before announcing his organization's objections to any federal control program. "Based on the mining industry's awareness of the economic factors involved," he said, "its experience in the diversity of the problem and the engineering techniques of land restoration, and its analysis of the problem on a national basis, the American Mining Congress is opposed to the legislation before you today."[13]

Opposition to each of the bills before the Senate committee also came from critics of surface coal mining. Harry Caudill testified at the hearings on behalf of AGSLP, the Congress of Appalachian Development (CAD), and the Sierra Club. The AGSLP, of course, had been established in 1965, in eastern Kentucky, with the intent of outlawing surface coal mining. CAD was founded a year later in Bristol, Virginia, with the dual purpose of promoting the public ownership and development of Appalachian resources as well as the construction of a series of new towns along the Appalachian mountain chain. The organization was the brainchild of Harry Caudill and Gordon Ebersole, a former assistant to Secretary Udall, who accompanied the Whitesburg, Kentucky, lawyer to the hearings as CAD's executive director. Caudill's connection to the Sierra Club dated back to a 1965 White House Conference on Natural Beauty, where he had a long discussion about strip mining with David Brower, the group's national president. Two years later, Lotts Creek Community School resident Johanna Henn formed a Cumberland section of the Sierra Club but found the state organization's officers unwilling to fight against surface coal mining. Caudill joined the new section after receiving a letter from Henn, and together they managed to get the state chair, James Kowalsky, interested in the issue. In December 1967, the Appalachian Volunteers and AGSLP sponsored a tour of eastern Kentucky surface mines for Sierra Club members, exposing them to the devastation there. But Henn and Caudill continued to experience difficulties getting the state organization to become actively involved in the campaign against stripping, and it was the national leadership who sponsored Caudill's trip to the hearings.[14]

Caudill opened his remarks to the Senate committee by linking environmental responsibility to American liberty. "Let us frankly recognize that the earth is just as important as the people who inhabit it and that the right to be free is matched by a responsibility to preserve freedom's land," he said. "Liberty in a wasteland is meaningless." Stripping should be allowed only where reclamation could be assured, he argued, and none of the bills before the committee met this standard. Adequate legislation would "outlaw strip mining in areas such as southern Appalachia where the slopes are so steep and the rainfall so great that reclamation and restoration of the land to its former utility is impractical and impossible." Since stripping and natural beauty were incompatible, Caudill maintained, it should also be banned in areas "of significant scenic loveliness and in important wildlife habitats." In addition, he expected any bill passed by Congress to include a massive government program for the purchase, reclamation, and revegetation of lands already stripped, financed out of a trust fund supported by a special levy on extractive industries.[15]

Other strip mining opponents were seemingly agreeable to the proposed bills. Alice Grossniklaus appeared before the Senate committee representing the Community Council for Reclamation (CCR), a group active in eastern Ohio since the early 1960s. CCR sought more effective enforcement of laws governing air pollution, water pollution, and restoration of land and property damaged by mining and other operations as well as enactment of additional regulatory laws. Grossniklaus was secretary of the group as well as president and owner of the Alpine Cheese Factory in Wilmot, the source of her interest in the situation for 1,500 dairy farm families in the area. She began her testimony with a presentation of photographs, the first of which showed part of Holmes County. "Nearly all land underladen with coal in the [eastern Ohio] counties affected are of this caliber," Grossniklaus explained, "productive, beautiful, filled with natural resources, near populated areas, or a combination of all. Truly God-given." She also showed postcards of the old-growth forest in the Stark Wilderness Center, which the council was instrumental in saving from strip mining. This visual presentation was followed by a listing of the reasons why good reclamation was imperative. In Meigs County, Grossniklaus reported, sportsmen were worried about the impact of stripping on their hunting and angling, the Agricultural Stabilization and Conservation Committee was concerned about the loss of income by family farmers due to stripping, and the county auditor claimed that a $150,000 decrease in valuation in the Scipio township was due primarily to surface operations there.[16]

National environmental organizations also lent support to the administration bill before the Senate committee. The National Wildlife Federation and the Conservation Foundation both accepted the cooperative federal-state regulatory approach embodied in S. 3132, although the latter suggested an amendment that would "prohibit any surface mining that would leave the surface less useful to man than it was before." Speaking for the Wilderness Society, John Hall endorsed the DOI bill as a way to address "the total disregard some surface operators have for the natural resources and for their fellow human beings." Representatives for the Izaak Walton League (IWL) made it clear, however, that federal involvement in regulating strip mining should leave as much authority to the states as possible. Assistant Conservation Director Roger Tippy said that the IWL membership favored a ban on some contour stripping as it was being practiced, but "we think that state actions in West Virginia, Kentucky, and Pennsylvania may be sufficient to deal with this contour problem." New, strong laws had just been passed, he said, and it was still too early to tell whether or not they would be sufficient. To prod other states that are ignoring the problems caused by strip mining, Tippy advised, the IWL supported the more "workable approach" of S. 3132, which allowed states to submit a control plan tailored to their needs rather than national standards. IWL executive director Grover Little Jr. followed with a declaration of his organization's opposition to federal regulatory action in the states where strong laws were in effect and enforced. "Our economic system is the best in the world but it sometimes exposes a questionable face," he said, "for it is paradoxical that we are destroying the beautiful mountains and valleys of one area to create an Eden in another." A federal law was necessary only as "an omen" to the states that had not yet established their own effective control program.[17]

Federal Legislation for a Ban

By the end of 1970, more and more people in the strip coalfields of Appalachia as well as outside the region were beginning to support a ban on surface coal mining. Indicative of this broadening support, in December the *New York Times* came out in favor of federal legislation to outlaw the practice, comparing strip mining to the dust storms of the 1930s. "[T]he Government today stands by, silently, impotently," wrote its editorial staff, "as coal operators lay waste the land and scatter the top soil as recklessly as the dust storms ever did." The explanation for the different responses to the two environmental crises, they contended, was "that nobody made money out of the dust storms." Yet not all of the new proponents of outlawing surface coal mining were truly committed to abolition, and this one factor had great implications for the whole movement. Very quickly after the campaign against stripping moved to the national level, it became dominated by representatives of organizations who strove to be "reasonable" and "flexible," viewing a ban only as a prod to get the House and Senate to pass strict controls. There were activists and even congressional representatives who had a principled commitment to outlawing stripping, at least in Appalachia. But by the end of the second round of hearings in 1971 and 1972, the efforts of the genuine abolitionists were being seriously undermined by the compromisers.[18]

The first federal bill to abolish surface coal mining was introduced by Democratic West Virginia representative Ken Hechler in mid-February 1971, just as the effort to ban stripping by state law was gaining momentum in Charleston. His proposed legislation, H.R. 4556, established an immediate ban on new operations, a six-month phaseout for strip mining begun before enactment, and the establishment of standards and procedures by the head of the Environmental Protection Agency (EPA) for state regulation of those operations. The bill also addressed problems associated with underground coal mining, directing the EPA administrator to draft "national environmental control standards" for all deep mine operations and prohibit underground coal mining that damaged wilderness areas or the resources of the national forest system. In addition, the proposed act included provisions for federal-state-local cooperation to reclaim abandoned surface and underground coal-mined lands as well as citizen suits against government officials for failure to implement sections of the law. And in May, Hechler's bill was amended by Ohio congressional representative John Seiberling to provide cash payments, counseling, training and placement services, relocation allowances, and priority employment on federally funded reclamation projects. This amendment was meant to insulate the proposal from charges that proponents of abolition were dismissive of the need for employment in areas where strip mining was conducted.[19]

Like Lausche's earlier study and control proposal, Congressman Hechler's abolition bill was a response to the demands of concerned citizens, largely from West Virginia and elsewhere in Appalachia. In January 1971, for instance, people living in the Big Creek District in McDowell County, West Virginia, sent him petitions, with more than 160 signatures, asking him "to take immediate and all necessary steps in halting surface mining operations in and around our communities and stopping the wholesale destruction of our forest land and wildlife sanctuaries."[20] But Hechler was already a committed opponent of strip mining in 1971, and his rhetoric evidenced a postwar populism. He contended that the administration's latest bill was "a milk-and-water approach," a "toothless law," and that it would be administered by an agency (the DOI) primarily interested in minerals production. This was not likely "to slow down the ruthless rape of the environment" by strip miners. Responding to the claim that abolition would cause unemployment, Hechler fervently replied:

> What about the jobs that will be lost if the strippers continue to ruin the tourist industry, wash away priceless topsoil, fill people's yards with the black muck which runs off from a strip mine, rip open the bellies of the hills and spill their guts in spoil-banks? This brutal and hideous contempt for valuable land is a far more serious threat to the economy than a few thousand jobs which are easily transferable into the construction industry, or to fill the sharp demand for workers in underground mines.[21]

If strip mining were abolished, the congressman noted, it would actually create more employment for deep miners. This would begin to deal with the loss of 300,000 jobs as a consequence of mechanization, which had caused little concern within the coal industry.

By mid-April, there were seventy-three House members from twenty-four states cosponsoring H.R. 4556, including representatives from surface mining states such as Ohio, Pennsylvania, Illinois, and Indiana. To add more cosponsors, Hechler organized a briefing for congressional staff members on April 22. From West Virginia he brought up Secretary of State John D. Rockefeller, House of Delegates member and retired miner Ivan White, underground miner and black lung activist Arnold Miller, and retired coal miner Clarence Pauley—all of whom presented the case against surface coal mining from their particular perspective. Ohio representative John Seiberling and Dr. Theodore Voneida, from Case Western Reserve University, also described the situation in their state. Partly as a result of this briefing, by the end of May there were eighty-seven cosponsors for H.R. 4556. And in the Senate, Gaylord Nelson (D-Wisconsin) and George McGovern (D-South Dakota) introduced S. 1498, an abolition bill that was nearly identical to Hechler's proposal.[22]

The introduction of abolition bills in Congress, and the increasing numbers of cosponsors for these bills, reflected the growing strength of the movement to outlaw surface coal mining. But even before Representative Hechler introduced his first bill, the threat of a ban had prompted the United Mine Workers and many of the larger coal companies and industry associations to support limited federal regulation. "State regulations have failed," UMW president Tony Boyle announced in January 1971, "and there has been lack of adequate standards and enforcement by the states." Contradicting much of what he had been saying just two years before, the union president argued that state laws required little in the way of restoration of stripped out land, enforcement was conspicuous by its absence, "and penalties are little more than a slap on the wrist." Union leaders were adamantly opposed, however, to the abolition of surface coal mining, which they described as "sheer nonsense," "so much political grandstanding," a "preservationist pipe-dream," and a threat to "badly needed jobs and essential electric power."[23]

Likewise, members of the coal industry changed their position because they were apprehensive about the growing effort to ban stripping. In 1968 the AMC had rejected all three of the bills considered by the Senate Committee on Interior and Insular Affairs. A year later, Consolidation Coal vice president James Riley had declared before the annual convention of the AMC that conservationists demanding better reclamation were "stupid idiots, socialists and commies who don't know what they are talking about. I think it is our bounden duty to knock them down and subject them to the ridicule they deserve." But in a special meeting in 1971, just before the beginning of new congressional hearings, the AMC Board of Directors adopted a statement of policy that broke with the remarks its representative had made at previous hearings. The AMC "will urge the adoption of realistic surface mining regulation at the state level and will support federal surface mining legislation which is realistically designed to assist the states and the surface mining industry in conducting surface mining operations," the statement said, "so as to have the least practicable adverse effect on other resource values." The AMC remained opposed to uniform national standards because of the diversity of conditions in mining areas and variations in methods, yet it was now willing to accept a role for the federal government in the regulation of strip mining.[24]

As the UMW and coal industry shifted positions, the opposition to stripping became even better organized at the regional and national levels. At an October 3 meeting in Huntington, West Virginia, representatives from Save Our Kentucky, Citizens to Abolish Strip Mining, Stop Ohio Stripping, the Wise County (Virginia) Environmental Council, and various individuals—

including Hechler, Galperin, and a staff member from Congressman John Seiberling's office—organized the Appalachian Coalition. They elected Jim Branscome as the coalition's coordinator, Richard Cartwright Austin as secretary, and various state representatives: Vicki Mattox for Kentucky, Ted Voneida for Ohio, Donald Askins for Virginia, and Gerald Sizemore for West Virginia. The coalition pledged to assist in the effort to ban coal strip mining through state and federal legislation, although following the failures to ban stripping in Kentucky and West Virginia it focused all its attention on enacting a federal law. Shortly after the formation of the regional group, possibly at an October 25 legislative strategy meeting at the Washington, D.C., office of Friends of the Earth (FOE), a number of national environmental groups organized the national Coalition Against Strip Mining (NCASM) and chose Louise Dunlap, on staff at FOE, as its coordinator. NCASM was not strictly abolitionist but would work in conjunction with the Appalachian Coalition to lobby members of Congress and arrange effective testimony at hearings.[25]

Also in October, Richard Cartwright Austin established the Appalachian Strip Mining Information Service (ASIS). At first, ASIS served members of only a few established groups in West Virginia, Ohio, Pennsylvania, and Kentucky, but the number of participating organizations quickly grew. The service's main component was a regular monthly bulletin, which was meant to provide timely summaries of information on the environmental, economic, and political aspects of strip mining as well as assist in coordinating activities. ASIS also provided access to a strip mine research library in the Charleston, West Virginia, office of Mid-Appalachian Environmental Services, where a full-time research staff was directed by Norman Williams, former deputy director of the West Virginia Department of Natural Resources. The library was initially funded by a $15,000 grant from the Conservation Foundation and it continued to function even after Austin ceased publishing the newsletter in 1972.[26]

While strip mining opponents were establishing new groups and creating new resources to enhance their influence at the national level, congressional committees began another round of hearings on surface mining legislation. Starting in September 1971, the House Subcommittee on Mines and Mining and the Senate Subcommittee on Minerals, Materials, and Fuels heard testimony from government officials, industry representatives, Appalachian activists, and spokespeople for national conservation groups. Speaking before the House subcommittee, Ohio congressional representative John Seiberling referred to problems with reclamation in the southeastern part of his state, particularly the irreparable damage done to the underground water system by blasting, but he also questioned the possibility of reclamation in the mountainous areas of Pennsylvania, eastern Kentucky, and West Virginia. The ideal solution from an environmental standpoint, Seiberling argued, was the abolition of surface mining. Deep mining could adequately meet the nation's coal needs and the comparatively small number of workers affected could be dealt with by his amendment to Hechler's abolition bill. Hedging on a permanent ban, however, Seiberling indicated that at the very least he favored a five-year moratorium accompanied by federal research and development.[27]

The lengthiest and most passionate remarks by a government official during the hearings were made by West Virginia congressional representative Ken Hechler. As Congress's leading proponent of outlawing strip mining, Hechler offered testimony to both the House and Senate subcommittees in support of his own bill and Senator Nelson's nearly identical proposal. He spoke before the House subcommittee for nearly three hours and faced a barrage of hostile questions and reprimands, many of which came from the chair of the full Committee on Interior and Insular Affairs, Wayne Aspinall, who did not usually attend subcommittee hearings. House subcommittee staff had advised Hechler that he could bring other witnesses to provide "expert" testimony on various subjects, and he brought Arnold Miller as well as forestry and economics professors from West Virginia University for this purpose. Mines and Mining chair Ed Edmondson made it immediately clear, however, that "direct testimony from these gentlemen will not be permitted at this stage in the hearings." Fortunately for Hechler, his appearance before the Senate subcommittee was not as difficult, and the Minerals, Materials, and Fuels chair allowed West Virginia delegate and retired deep miner Ivan White to add comments at the end of the scheduled testimony.[28]

Speaking to the House subcommittee, Congressman Hechler attacked surface coal mining from a number of angles, but he characterized stripping generally as a detriment to both the landscape and the economy. "Representing the largest coal-producing State in the Nation," he said, "I can testify that strip mining has ripped the guts out of our mountains, polluted our streams with acid and silt, uprooted our trees and forests, devastated the land, seriously destroyed wildlife habitat, left miles of ugly highwalls, ruined the water supply in many areas, and left a trail of utter despair for many honest and hard-working people." Hechler also reiterated what he had said during the earlier hearings about surface coal mining and the Appalachian economy. Stripping was destructive of jobs, both the jobs lost when stripping replaced deep mining (a process which he saw as part of a larger problem of technological unemployment in the coal industry) and the employment lost when destruction of the landscape made an entire area unattractive for tourism. The abolition of surface coal mining, on the other hand, would necessitate the employment of thousands of additional underground miners and still allow for a tourist industry based on scenic value. Members of the House subcommittee challenged Hechler on his argument that strip mining jobs were temporary, pointing out that deep mining jobs were temporary too, and they forced him to admit that while tourism employed more people than stripping, its jobs paid much less. But in fairness to Hechler, these arguments were disingenuous coming from representatives who were attempting to defend the coal industry, which had cut its workforce by nearly 70 percent since the end of World War II.[29]

Probing further, Congressman Edmondson questioned Hechler on the difference between deep and strip mining in terms of safety as well as economic efficiency, and this led the West Virginia congressman into an involved comparison of the two methods of coal extraction. Fatalities occurred in both types of mining, he declared, and the safety record of strip mining was not much better than the record of deep mining. Comparing the two on the basis of deaths per millions of man-hours of exposure, in 1970 there were .66 deaths at strip and auger mines and 1.17 deaths at deep mines. On the issue of black lung, a respiratory disease caused by inhalation of coal dust, Hechler claimed that the Coal Mining Health and Safety Act, which he shepherded through Congress in 1969, would reduce its incidence and make it so that "young miners will have little to worry about." Addressing the problem of subsidence of land above deep mines, he argued that it occurred only in areas unregulated by law, and that when mining was uniform and pillar strength was adequate, subsidence was negligible. Stripping, however, affected the entire land areas above a coal seam as well as areas off the site. With regard to acid mine drainage, deep mines could be purposely collapsed, flooded, sealed, or back-filled as preventive measures, whereas controlling the movement of acidic spoil and groundwater at surface mines was much more difficult. Aesthetically, deep mining and stripping were not comparable, Hechler declared, and strip mining destroyed the productivity and usefulness of the land, depressed appraised land values, and eroded the tax base, "[and its] effects on the environment discourage location of other industries, commerce, [and] housing."[30]

Members of both the House and Senate subcommittees also questioned Hechler about his bill's impact on the nation's ability to meet its energy requirements, an issue that would eventually create an important split within the opposition to surface coal mining. Hechler suggested that the nation's coal supply would be adequate with the opening of additional deep mines, which he said might take from one to three years at the most. "And what are we going to do for the six months to thirty months time lapse," asked Congressman McClure, "between the closure of the surface mines and the [new] production from underground mines?" Hechler responded by pointing out a built-in time lag in the legislative process and suggested a reduction of coal exports (which had increased from 16.5 million tons in 1940 to 70.9 million tons in 1970). He wanted to have Arnold Miller speak at this point, to explain the possibility of adding more shifts at existing mines on a temporary basis and talk about working in a mine on Cabin Creek that had taken a year to open, but the subcommittee would not allow it. "It seems to me that we are approaching the problem all wrong," Hechler finished in exasperation, "by saying we have an energy crisis that we are caught in. . . . So we are going to have to continue to rip up the land and pollute the water."[31]

Testimony before the House and Senate subcommittees by industry representatives revealed the important shift in their position on regulation. Coal industry officials endorsed limited federal-state cooperation, granting the federal government the authority to establish broad regulatory standards that would allow for "flexibility" and leaving it up to state agencies to enforce them. "From State to State, from place to place, it can well be said of mining that its only constant is its diversity," argued American Mining Congress chair Joseph Abdnor. "All such diverse realities of mining argue eloquently against any effort to devise other than broad, reasonable Federal guidelines." And since state authorities were most familiar with their particular terrain and climate, they were most capable of implementing the federal standards. But industry representatives also made a point to argue against outlawing surface coal mining. Reclamation was possible, said National Coal Association (NCA) President Carl Bagge, whereas prohibition "would have disastrous results for the Nation and its constantly increasing need for energy." Bagge argued that it was unrealistic to expect that surface-mined coal could be replaced by deep-mined coal, which would require 132 additional underground mines of 2 million tons annual capacity, a capital investment of $3.2 to $3.7 billion, as well as 78,000 additional trained deep miners. It would also take three to five years, he claimed, before the mines reached full production. Even prohibition at the state level or in specific areas was unwise because mining and reclamation "which is impractical in some areas now maybe quite feasible next year because of new developments in technology."[32]

In their testimony at the hearings, UMW officials expressed both a desire to have economic growth and a healthy environment as well as fear of a ban if federal regulatory action did not come soon. They voiced support for H.R. 10758 and S. 2777, which would require coal operators to post performance bonds, apply for federal permits, and establish an advisory commission with representatives from the EPA, Department of Agriculture, and Department of Interior to administer the law. Speaking before the House subcommittee, Joseph Brennan explained that his union had an interest in surface mining regulation because miners and their families lived near strip mines, earned their living from strip mining, "and a great deal of revenue [from strip operations] goes to the UMWA Welfare and Retirement Fund and the Anthracite Health and Welfare Fund." But the union was primarily worried that under-regulated strip mining would strengthen the abolition movement. Leonard J. Pnakovich, president of District 31 (West Virginia), explained, "We know that to the extent the surface mining industry continues to devastate our landscape the jobs of coal miners are in jeopardy." Later in 1971, speaking before the Senate subcommittee, Brennan warned, "Continued abuse of America's precious land and water resources because of unregulated strip mining must ultimately lead to a citizen revolt against all strip mining." And at year's end, President Boyle summed up the campaign the union had led for controls on surface coal mining, reiterating much of what Brennan and Pnakovich said before Congress. The abuse of land resulting from surface mining was both inexcusable and unnecessary, he said, yet "the cure should not be to forego further surface mining, but rather, to stop the adverse effect which it causes."[33]

The House and Senate subcommittees also heard from a number of Appalachian activists, most of whom emphasized the social impact of strip mining and demanded that Congress pass an abolition bill. Save Our Kentucky director and Appalachian Coalition chair James Branscome argued that passing a regulatory bill would be "overlooking the fact that the environmental damage is not nearly so great from strip mining as it is an affront to human welfare, property rights, and the apolitical process in the coalfields." Strip mining in Appalachia was the cause of unemployment, a decreased tax base, and massive out-migration, he said, and with the exception of Ken Hechler, the region's people had no one voicing their concerns in Congress. Efforts to deal with surface coal mining through controls also ignored the failure of the most stringent reclamation laws. "It is obvious to anyone who does not see with the eyes of greed," Branscome explained, "that a scraggly locust plant is not a grand oak, that a silt dam is not a protector of pure streams." People had sought redress for these social and environmental grievances through the courts and legislatures, he noted, but they had "learned that the only order is that which protects the strip miners." Branscome explained that his appearance before the committees was an effort "to get the legal processes to work before the people have to help themselves." Residents of the hills and hollows were patriotic, God-fearing, patient, and cautious, but the strip miners were making revolutionaries out of them. People were telling Branscome they were going to start taking up weapons if the political process failed them once more.[34]

Following Branscome's testimony before the House subcommittee, SOK chair Vicki Mattox delivered a personal account of the damage done by stripping. Mattox was the daughter of a Letcher County, Kentucky, deep miner and her interest in speaking to the subcommittee was to supplement the many comments witnesses were making about the impact strip mining had on the environment with an explanation of the effects it had on people. Most of the people she was concerned about could not afford to travel to Washington, D.C., to testify for themselves, she said; some could not read or write, but their stories were no less important than those of other "expert" witnesses. Mattox told about one "old gentleman" on Yellow Creek, in Knott County, Kentucky, a fifty-year resident of a stone house he had built with his own hands. Acid drainage from strip mining at the head of the hollow had turned the creek red and erosion had filled it with silt, ruining the man's water supply. For a time, a neighbor brought the old man water—otherwise he would have had to move—but when Mattox visited last summer the home was deserted and surrounded by weeds. This story was a common one, she implied, and all the problems apparent from a walk along the seemingly endless miles of strip mine benches suggested that "Kentucky's strict reclamation laws are powerless to stop the destruction and human suffering."[35]

Focusing on the conditions in Tennessee, J. W. Bradley further hinted at the revolutionary potential of the fight against strip mining. Bradley was the son of a coal miner, had himself been a coal miner for seven years, was currently working as an electrician, and had traveled to the hearings at his own expense. He did not represent any particular organization at the time, although very soon after the hearings he assumed leadership of a new group in Tennessee called Save Our Cumberland Mountains (SOCM). "I am for stopping strip or surface mining as soon as possible," Bradley explained, because stripped land could not be reclaimed and strip miners took the jobs of deep miners. He also pointed out to subcommittee members that the Declaration of Independence said whenever any form of government was destructive to the ends of protecting the peoples' inalienable rights, it was the right of the people to alter or abolish it and institute new government. "I am not advocating that we overthrow the Government," Bradley clarified, "but I think it is time they check into the situation, and see that the things that I have stated are factual and also see if [government officials] can do something about it." Another Tennessee resident, Robert Peelle, spoke for the Tennessee Citizens for Wilderness Planning (TCWP), a statewide organization of five hundred members. Peelle explained that his group was in favor of a federal control law to create uniform regulation across state lines. TCWP urged "complete and rapid prohibition of surface mining for coal as we know it now in certain environmentally and socially defined circumstances," and strong regulation elsewhere, "to assure the full restoration of surface values to what they are at present."[36]

Representatives of national conservation groups, including the Friends of the Earth, Sierra Club, Wilderness Society, National Wildlife Federation, Conservation Federation, Izaak Walton League, National Audubon Society, and the Natural Resources Defense Council, declared their organizations in favor of a ban, but they hedged on this support and spelled out their interests in federal regulation too. Louise Dunlap, the assistant legislative director at FOE and director of NCASM, argued that H.R. 4556 was most consistent with the National Environmental Policy Act, the Coal Mine Health and Safety Act, and "the integrity of the free enterprise system which cannot survive with continued cost of operations being passed on to the taxpayer as social cost rather than to the consumer." In a letter to the Senate subcommittee, Dunlap also expressed her agreement with the assumption of S. 1498 that restoration of the land could be achieved only under rare circumstances, and she pointed out that the bill offered an increased dependency on deep mining as an alternative to stripping. She did not deny that reclamation could ever be achieved, but argued that given the state of the art of reclamation and the political and economic pressures on the strip mine industry to not fully internalize all operating costs, the focus should be on how to preserve the surface for future generations.[37]

Like the other representatives from national conservation groups, however, Dunlap outlined what needed to be changed in various regulatory bills if Congress chose not to outlaw stripping. She suggested that the maintenance of maximum ecological value should be made "the" rather than "a" prime consideration in H.R. 10758. Although there was broad-based support for Nelson's abolition bill, she also told the Senate subcommittee, Senate action "which attempts to regulate rather than abolish strip mining should prohibit strip mining under certain conditions and give the EPA administrator authority to designate areas unsuitable to strip prior to granting any new permits in an area under consideration." Months later, after moving to the Environmental Policy Center (EPC) but retaining her position at NCASM, Dunlap wrote to Representative John Saylor requesting his vote for H.R. 6482. The bill fell short of a total phaseout of surface coal mining, she acknowledged, and it was inadequate in its failure to give primary federal authority to the Environmental Protection Agency, "but it does provide an excellent test for regulation." In closing her letter, she warned that failure to enact the bill would "confirm growing public opinion that the only way to regulate coal surface mining was to ban it."[38]

Employing much stronger rhetoric than Dunlap, Sierra Club eastern representative Peter Borelli described the state laws designed to control stripping as weak attempts "to placate the public conscience through a series of loose regulations." In Pennsylvania, the toughest and best enforced of the state laws had brought noticeable improvements, but even there "spoils are still unstable, the slopes still erode, acid still leaches into the streams, and the consecutive ridges of area stripping are still as ugly and useless as ever. Black locusts and legumes still struggle to provide at best spotty growth over the barren and poisonous mounds of pulverized rock and shale." Likewise, although TVA adopted "a very good contract provision" on reclamation, it was not being enforced by the agency. The failure of state laws and the TVA contract clause, Borelli argued, suggested that the role of the federal government should be prohibition of strip mining rather than its regulation. In May 1970 the Sierra Club board of directors had voted for "a total and immediate ban on all surface mining of coal," and Borelli communicated the organization's support for H.R. 4556 and S. 1498 to the two subcommittees. But in May of the next year, his report in the *Sierra Club Bulletin* exposed the organization's strategy. "Though no one seriously expects [the abolition bills] to win committee approval," Borelli wrote, "the abolitionists and their legions have been the first to budge these traditionally mineral-oriented committees. As a result several milder but potentially effective measures that might otherwise have been ignored have earned some credibility."[39]

Given the opposition of both coal industry representatives and mine worker officials to a ban, as well as the vacillation of national environmental organizations on abolition, Appalachian opponents of strip mining had some reason to feel discouraged about passage of an abolition bill and signs of frustration soon became evident. On December 4, 1971, more than two hundred activists from West Virginia, Kentucky, Virginia, Tennessee, and Ohio gathered in Wise, Virginia, for what was dubbed a "People's Hearing on Strip Mining." It was difficult for the people of Appalachia to speak collectively and be heard, explained Warren Wright in his introductory remarks to the meeting, because they had no advocates, no churchmen, and no elected representatives on their side, with the exception of Ken Hechler. "Of late we have been attempting to speak before other tribunals—in Washington and Frankfort and Charleston, and soon perhaps Richmond—before senators, Representatives, Reclamation officials—and the consensus I seem to gather from our homecoming people is this—'They tolerated us.'" Opponents of strip mining needed to be heard demanding the things that are written into the so-called American guarantees, Wright said. "We must lay hold of the justice that for the poor men is now existing only in the pages of law libraries . . . we are our own political salvation." Other speakers echoed Wright, including John Tiller from Brammel, Virginia, who acknowledged the pointlessness of "going to Washington and seeing this little punk congressman or senator, who's already been bought and paid for many years ago." Alma Cornell, of Wise, bemoaned the fact that the nation's laws were no longer for the people, by the people, or of the people, but only for a few. "Surely the people deserve laws to provide protection for their property, health, home, and safety of lives against a few who do not care how much damage they do to others, to make millions for themselves."[40]

Such frustrations gave rise to renewed civil disobedience when severe flash flooding in April 1972 killed one person and damaged many homes, gardens, bridges, and roads in Floyd County, Kentucky. Some local residents thought the flooding was caused by heavy runoff from a local strip mine, and approximately thirty of them held several meetings to consider what form their response would take. After one of the meetings they began circulating petitions to the state department of reclamation to ban strip mining in the area, and they made it clear that if the petitions did not work they would take other action to close down the mine. The flood victims collected nearly a thousand signatures on the petitions and sent them along with a representative to the Kentucky Division of Reclamation. Nothing was done by the agency, however, and the director later denied receiving any complaints. Then, in late June, two hundred Floyd County residents forced the offending strip operation to temporarily shut down. One man walked up to a bulldozer operator to tell him to stop the job. When the operator kept the machine running, the rest of the crowd walked up, the operator turned the bulldozer off and got down, and all other work at the site stopped for the day.[41]

Also in June, over nine hundred activists gathered at a National Conference on Strip Mining at the Union College Environmental Center in Cumberland Gap National Park, near Middlesboro, Kentucky. Called by Senator

Fred Harris (D-Oklahoma), the meeting was meant to create a truly national coalition of surface coal mining opponents as well as draft a comprehensive statement on stripping to be included in the platforms at the Democratic and Republican conventions later that year. Participants at the meeting read like a "who's who" list of people involved in trying to ban or regulate stripping since the late 1950s, including both grassroots activists and leaders of national conservation groups. But also present were representatives of the coal industry. This mix of participants kept debate lively. Much of the day's discussion centered on the corporate exploitation of natural resources for profit and the need to tie abolition together with job creation. Responding to the likes of Elkhorn Coal's Paul Patton, who said that strip mining opponents cared only about aesthetic value, Helen Wise acknowledged that strip mining "put bread on the table for a few," but argued that it also destroyed "the homeland—not only for [strip mine employees] but for all their kin and for all their neighbors." There was certainly a need for federal programs, she said, "to develop the land—develop good jobs for people, develop a way of making a living and staying in the mountains without destroying the mountains."[42]

By the end of the conference, participants had produced a draft of a resolution, which was adopted by the national Coalition Against Strip Mining two days later in Washington, D.C., and presented to the platform committee of the Democratic National Committee in Denver the day after that. Surface coal mining was attended by abuses of land, water, and people, the resolution read, and the social costs caused by the particular method of resource extraction were not only going unpaid but were irreparable. Other methods of coal mining could fully compensate for the loss of tonnage caused by a ban on stripping. "THEREFORE, be it hereby resolved that STRIP MINING of coal in these United States be henceforth abolished," the resolution stated, "with no new permits issued from the date of enactment [of an abolition law], allowing for six months phase-out during which contour strip mining operations would close and two years phase-out during which current area strip mining operations would close." The resolution also included a ban on federal purchasing of strip-mined coal, outlined reclamation requirements for current stripping operations, and called for primary federal authority during the phase-out period to be given to the EPA. All these provisions were included in Ken Hechler's abolition bill—though there was a difference in the phase-out timetable for area stripping—and many of the groups favoring H.R. 4556 were signatories to the resolution. However, the Democratic

National Committee responded by endorsing only general opposition to strip mining.[43]

By July 1972, the House and Senate subcommittees had reported bills to their full committees, and Louise Dunlap made assessments of the measures that reflected her interest in passing a regulatory bill rather than outlawing stripping altogether. The Senate bill, S. 630, would probably receive widespread coal industry support, she said, because it was so weak. The bill covered all minerals and included specific reference to coal in only several sections, gave federal authority to the Department of Interior, allowed operators to "recondition" rather than reclaim a site (and left "recondition" undefined), gave states up to two years to begin enforcement of federal standards, made few demands on states with regard to monitoring and inspection, and failed to include a provision for either notifying or getting permission from surface owners. The Senate proposal also allowed operators to leave highwalls and benches, which had been proven to cause landslides and flooding, Dunlap noted, and it allowed them to construct water impoundments similar to the one that broke and sent a massive, deadly wall of water through the valley of Buffalo Creek, West Virginia, in February. No bill would be better than the Senate bill, she suggested, because S. 630 weakened the regulatory progress made in many states over the past decades.[44]

The House bill, H.R. 6482, was not an abolition bill either, but it was better. "If one is willing to accept that strip mining can be regulated and should be allowed to continue," Dunlap advised, "the House bill would be considered a well constructed bill." The measure was "threaded" with amendments made by John Saylor to make it nearly as stringent as the Pennsylvania law, but also included counter-amendments made by Wayne Aspinall to limit public notification, participation, and appeals as well as to minimize specific criteria to designate areas unsuitable to strip. "The results of the interplay to strengthen and weaken the bill," she explained, "provide us with a bill which has clarified its legislative intent, while eliminating a number of critical specific provisions which would require that the intent be carried out, rather than merely suggesting that it be done." In September both the Senate and House committees reported out versions of S. 630 and H.R. 6482, little changed from the bills the subcommittees sent to them.[45]

As the regulatory bills passed through subcommittees and committees to the full House and Senate, the United Mine Workers underwent organizational changes that had important implications for their position on surface coal mining. President Boyle had held his position against Yablonski in 1969 voting fraud and murder. But rather than weaken the democratic insurgency, Yablonski's death consolidated the reform movement. A new organization, Miners for Democracy (MFD), was born at his funeral in spring 1970. With the intention of challenging the Boyle regime again in the 1972 elections, MFD founders held a convention in Wheeling, West Virginia, in May 1972, to choose a slate of candidates and write a platform. For their presidential candidate, the reformers chose Arnold Miller, a deep miner for twenty-five years from West Virginia and a leader in the black lung movement. During the previous year, he had participated in a panel discussion at a meeting of Citizens to Abolish Strip Mining in Madison, West Virginia, and called for a "massive demonstration" to push Si Galperin's abolition bill through the state legislature. Miller also appeared with Ken Hechler before the House subcommittee, in September 1971, to lend support to the West Virginia representative's testimony for a ban on surface coal mining. Yet he backed away from an abolition position in 1972. The MFD platform addressed "the surface mine controversy" with a plank that was largely the creation of Bill Kelley, president of Local Union 7690, a surface mine local in eastern Ohio. This plank did not propose a ban, stressing instead "BOTH jobs and land." Nonunion strip miners were to be unionized and reclamation laws enforced. As Miller later explained, the plank said "that wherever surface mining was done with responsibility, we approved of that, and wherever it wasn't we didn't approve."[46]

During his campaign for the union presidency, Miller suggested that under his leadership the UMW would organize the unorganized, particularly surface miners, and work for good reclamation. Nonunion strip miners were producing millions of tons a year, he noted, and no royalty was paid to the Welfare and Retirement Fund on any of this coal. "When I take office we are going to launch a full-scale peaceful drive to bring these men into the union," Miller promised in an open letter to a supporter, "resulting in a stronger UMW and millions of dollars a year in additional royalties for the Fund." The MFD slate also ran a full-page campaign advertisement in the *United Mine Workers Journal* with a banner headline reading, "Bill Kelley Thinks It's Time For a Change—So Do a Lot of UMWA Strip Miners." In the ad Kelley noted the widespread support for "tough laws to restore land to productive use." He had worked for this legislation at the state level, and the MFD slate had "taken a strong stand for restoration done under UMWA contract by UMWA members." Put this way, regulatory legislation needed to reverse the effects of surface mining would create jobs, possibly union jobs, rather than take them away. "We live where the coal is mined," Kelley added, "and we're the ones with the most at stake if we don't have good laws and a strong union to see that they get enforced."[47]

Miners for Democracy was victorious in the 1972 election at both the national and district levels. Yet Miller just barely won the presidency with 55 percent of all votes cast. He earned a majority in only six of the ten districts in Appalachia.[48] MFD had scored a major victory, the first ever of a rank-and-file slate, and it opened up the opportunity for transforming the United Mine Workers into a democratic, militant, and powerful organization. But the slim margins that carried the reform candidates into office indicated that any transformation would be difficult. Later, in the few years after the election, divisions within the union created a context in which the UMW's position on strip mining regulation could be weakened. Likewise, as early as 1972, divisions among opponents of surface coal mining were beginning to threaten the movement to outlaw the practice through federal legislation.

Getting More and More Cynical

Decline of the Opposition, 1973–1977

Radical opposition to strip mining was strong from the mid-1960s to the early 1970s, buoyed in part by other social activism of the time, including a more broad-based environmental movement. Very soon after proponents of a ban started to focus on passing a federal law, however, the movement began to weaken and dissipate in the face of numerous challenges to its efficacy. After 1972, with a few notable exceptions, representatives of local and regional groups dedicated to outlawing strip mining seemed to be increasingly willing to compromise and accept strict national standards and federal-state enforcement. Yet as an abolition bill became less central to the opposition movement, the regulatory bills introduced by members of the House and Senate became weaker. With the pressure off, members of Congress who were initially reluctant to support control legislation pared down others' proposals or introduced watered-down measures of their own.

The factors responsible for the decline of abolition sentiment were varied. By the early part of the decade, government-sponsored antipoverty work had all but ceased in Appalachia and, to the extent Appalachian Volunteers and VISTA workers helped galvanize opposition to strip-mining, their absence hurt the grassroots movement. Coal operators had a more direct and active role in the decline. Replicating a strategy that had been successful at the state level, the operators relied heavily on their close ties to federal government officials and dissuaded senators, representatives, agency bureaucrats, and even presidents from doing anything that would hurt the industry. Contributing to their success in applying this pressure, the economic downturn and a so-called energy crisis of the 1970s allowed coal industry representatives to more effectively present the debate over strip mining as a mutually exclusive choice between jobs or the environment. This argument also helped to undermine the United Mine Workers' short-lived attempt to take a more independent approach to the controversy. Once in league with coal operators in their opposition to regulations, after 1972 UMW leaders had struggled to find a balance between the union's (short-term) economic interests and the integrity of the natural environment. During the second half of the 1970s, however, divisions within the UMW intensified and this led the miners' organization to support weak, state-level legislation on the eve of the enactment of a federal law. Finally, and perhaps most significantly, radical opponents of strip mining were ill-served by national environmental leaders who complemented the rhetoric of the coal industry by insisting on the need to be "realistic." The middle-class, professional representatives of organizations like the Environmental Policy Center and the Sierra Club helped create a legislative context inhospitable to federal abolition. By mid-decade, they had convinced many strip mining opponents with deeper roots in Appalachia to lend their influence and resources to passage of regulatory legislation rather than an abolition bill.

House and Senate Hearings, 1973

In 1973, strip coal operators still considered the movement to outlaw stripping a threat to their business and they continued to support a role for the federal government in regulating surface mining, although they balked at anything more than nominal regulatory standards. Operators' interest in staving off prohibition by support for federal control legislation was evident in the testimony industry representatives gave at a new round of congressional hearings, held before a Senate committee and House subcommittees between March and May. "By any yardstick of reason," WVSM(R)A president James Wilkinson told the Senate committee, "those who advocate the elimination of surface mining for environmental protection could only be interpreted as ill-advised and unrealistic. . . . At best, it is an extremist solution to what is essentially an esthetic problem." The loss of surface mine production could not be made up by deep mining, and outlawing stripping would do serious damage to the economy of West Virginia and the nation, he said, but the wvsm(r)a was interested in a federal law that equalized regulatory standards in the states. The strip mining industry in West Virginia, Wilkinson noted, "supports comprehensive legislation establishing criteria for sound reclamation and requiring States to enforce regulations to meet Federal standards." Yet it was not entirely clear what sort of state-level standards coal organizations were willing to accept. Citing "differences in terrain from State to State," Wilkinson rejected specific limitations on the height of highwalls, degree of slope, and bench width.[1]

Leaders of national coal industry organizations also opposed any degree of prohibition while simultaneously acknowledging a willingness to accept at least vague federal standards administered as part of state programs. Speaking to House subcommittees for the American Mining Congress and National Coal Association, Peabody Coal president Edwin Phelps called the abolition of surface mining "unrealistic and irresponsible, not only because 50 per cent of U.S. coal production is mined by surface methods, but because it ignores the fact that the technology exists for the effective reclamation of mined lands." The coal industry supported only "realistic federal legislation," which he defined as "the approach which encourages the states to develop their own programs based on broad federal criteria." Climate, soil, vegetation, and topography differed greatly throughout the country, Phelps said, and these local conditions were an argument against specific, uniform reclamation requirements as well as for the administration of a federal statute by state officials, who knew best how to cope with the problems in their particular area. Speaking before the Senate committee, Phelps also took the opportunity to suggest that any increased cost in mining coal would be passed on to utilities and then to the "consumers and voters of America."[2]

The United Mine Workers made their interests known to members of Congress during the new round of hearings too. As a result of the election of the reform slate in 1972, however, the union and coal industry expressed different positions on the regulation of surface coal mining. Soon after taking office, umw president Arnold Miller forwarded a statement to the Senate committee in support of strong federal regulation as well as selective abolition. This statement reflected a sincere personal interest in the matter and presented the umw as an organization that would insist on good reclamation, but it also revealed conflicting views within the union. Miller argued that surface mining should be allowed where it is the only feasible way of getting coal out (such as in the Pennsylvania anthracite field) and only where the land could be restored. He also called the bluff of the coal industry and energy conglomerates, which had been arguing against strict regulation on the grounds of an energy crisis, an argument old-guard UMW leaders had used as well. Yet, while Miller expressed the UMW's hope that legislation passed by Congress limiting surface mining would protect displaced workers, he declined to endorse any of the pending bills. The UMW International executive board had given him the authority to speak for restoration of the land and assistance to surface mine employees affected by regulation, but that mandate did not include supporting a specific piece of legislation.[3]

In his own testimony before the House and Senate committees, Representative Ken Hechler continued to defend the need to outlaw surface coal mining. He had modified his abolition proposal somewhat, keeping the six-month phaseout for contour strip mining but extending the phase-out period for area stripping to eighteen months, adding more detailed provisions for the regulation of strip mining during the interim, and incorporating Seiberling's amendments to provide assistance to displaced workers. Yet the changes Hechler made in his bill were in no way a sign of lagging fervor. In remarks to both congressional committees, the West Virginia representative began by announcing that a revolution was brewing in Appalachia. People were not going to stand by while strippers ripped up their homeland, he said, and if Congress passed an innocuous bill designed to quiet public outcry while meeting the demands of the coal industry, "then there'll be a Boston Tea Party which won't be a tea party." Speaking before members of the House subcommittees, however, Hechler took pains to show that he was not being "emotional," as critics had begun to label prohibition advocates. "I am dealing here in hard and realistic facts and figures which relate to the energy demands of the Nation and how we are to meet these demands," he said, "while conserving the essential and limited resources which we possess in a world troubled by a war as a result of which the Arab Nations have imposed an embargo on oil shipments." Hechler then went on to make a detailed comparison of underground and surface coal mining methods similar to the one he introduced in 1971, emphasizing the greater amount of deep-minable, low-sulfur reserves and the prospects for safer and more efficient underground mines.[4]

Individually and organizationally, activists supported Hechler's new bill, but the foundation of the opposition movement was crumbling by the spring of 1973. Just before the Senate hearings, Appalachian Coalition president Reverend Baldwin Lloyd and Virginia coordinator Jim Coen called on fellow group members to write their senators and representatives in support of H.R. 1000, if not to achieve a legislative ban then to get better regulation.

"Even if this bill is beyond our hopes to attain," they explained in a missive to members, "it does serve as a spearhead for strong regulatory measures in subsequent bills." The two leaders explained that Louise Dunlap, head of the national Coalition Against Strip Mining and on staff at the Environmental Policy Center, had been drafting such legislation. But they lamented the "loose and undefined" structure of the Appalachian Coalition, bemoaning the fact that they—the leadership—"have had no response from much of our constituency and we don't really know where we stand in terms of solidarity." Later, in July, Lloyd wrote to Representative Hechler claiming to speak for NCASM and declaring support for a "strong regulatory bill" that prohibited surface coal mining on slopes exceeding 14 degrees, required surface owner consent, granted the right of public hearings prior to permits being issued, and allowed for citizen suits to ensure law enforcement.[5]

In addition to being president of the Appalachian Coalition, Reverend Lloyd was also one of a number of individuals who articulated a Judeo-Christian argument for limited abolition and strong regulation, a position that became more prevalent during the new round of hearings. Lloyd had settled in Blacksburg, Virginia, in 1958, working as an Episcopal priest and as executive director of the Appalachian People's Service Organization (APSO), an Anglican Appalachian outreach agency. Later he directed the Operation Coal Mining Project of the Commission on Religion in Appalachia (CORA), a cooperative effort of seventeen denominations concerned about the region. Like other outsiders, Lloyd worried about what strip mining was doing to the "fiercely proud, independent highlanders," but immersed as he was in faith-based communities, he also saw the controversy over strip mining in Appalachia as a moral issue. When Lloyd appeared before the House subcommittees, he presented the statement of concern CORA adopted a year earlier, which laid out the "assumptions" underlying the organization's support for at least partial abolition in mountainous areas. All resources of the earth belonged to God, the statement declared, and their "use must at all times be governed by His laws and guided by His purposes." Private ownership of property was a man-made institution, which needed to be "disciplined and guided by the over-arching responsibility of stewardship." Resources were meant to be developed and used, the statement continued, but a fundamental aspect of stewardship was the preservation of the "remarkable ecological balance" contrived by God.[6]

Other religious leaders and organizations also contributed to the development of a moral argument against surface coal mining. Speaking for the Church of Brethren Appalachian Caucus, former antipoverty worker Michael

Clark told the House subcommittee members that the violence against the land and people represented by stripping was in conflict with the nonviolence "of our religious heritage." No amount of profit could justify the practice, he said, and it should be banned in Appalachia. In a policy statement forwarded by Stephen Bossi, the Washington, D.C., representative of the National Catholic Rural Life Conference, his organization also went on record in support of outlawing contour mining and strictly regulating area stripping. Energy needs did not justify the destruction of productive land and the endangering of human life, the statement explained, since deep mining and renewable resources were not being fully exploited. In a letter to Pennsylvania Congressman John P. Saylor, sent on behalf of the National Catholic Rural Life Conference and a slew of other denominational organizations, Bossi outlined additional faith-based objections and recommendations. As CORA and other religious spokespeople had done, Bossi quoted the first verse of Psalm 24, which declared that the earth belonged to the Lord and implied that those who "dwelt therein" had a duty of stewardship. "The unnecessary despoiling of our land and endangering of human life through strip mining," he argued, did not reflect the respect for creation that Psalm 24:1 and other Biblical passages seemed to demand. "At the very least," Bossi explained to Saylor, "your committee should prohibit contour strip mining on slopes greater than fourteen degrees," which would follow consistently from congressional representatives' declared support for prohibition in areas that could not be reclaimed.[7]

Judging by hearing testimony, most of those activists with deep roots in Appalachia were still pressing for quick, national abolition. Blackey, Kentucky, resident Joe Begley warned that violence could not be ruled out in the fight to stop stripping and then recommended its ban by federal law. Council of Southern Mountains representative Betty Jean Justice, from Clintwood, Virginia, declared herself in favor of abolition, "because my experience with regulations is once those regulations are overstepped there is no adequate remedy." Similarly, J. W. Bradley, a former deep miner in eastern Tennessee and leader of the newly organized Save Our Cumberland Mountains, pointed out the failures activists had with regulatory bills and state legislatures and insisted that the federal government had the responsibility "to see that strip mining be abolished as soon as possible." Harlan County, Kentucky, resident and Black Lung Association member William Worthington pointed out strip mining's devastating impact on local employment, property, drinking water, watersheds, and wild game, and called for a ban on the practice. "I am for abolishing strip mining," he said, "because it does infringe upon the rights of others." Yet a notable few mountain residents seemed increasingly willing to achieve a partial ban through slope limitations. In an early indication of wavering resolve, longtime Hazard, Kentucky, activist and retired deep miner Mart Shepard outlined ways to improve contour strip mining and suggested a 12-degree slope limitation. This partial ban would have effectively stopped stripping in much of the mountainous parts of Appalachia, but it backed away from the forceful and determined insistence on the complete rejection of surface coal mining that drove the militant opposition.[8]

Those with more shallow roots in Appalachia tended to hedge even more on the demand that Congress outlaw stripping. By 1973, Richard Cartwright Austin was co-chair of the national Coalition Against Strip Mining and, speaking before the House subcommittees, he voiced support for H.R. 1000, "because it approaches this fundamental problem most directly." But Austin was quick to acknowledge that it was unlikely the hearings would result in legislation to abolish or curtail (partially ban) surface coal mining. "In the event that this committee attempts regulation," he said, "I must make one more point on behalf of the people I represent and the lands that they love." A strip mine control bill, Austin explained, should restrict surface coal mining to areas where land and waters could be restored, include a clause requiring surface owner consent, and allow for public scrutiny of stripping (through open hearings on proposed mining, reclamation plans, and assessments of whether or not reclamation standards have been met). Questioned by the subcommittees' members about his presentation of two different recommendations, Austin described the latter part of his testimony as "a fall-back position."[9]

Louise Dunlap, the other co-chair of NCASM and the Washington, D.C., representative for the Environmental Policy Center, also testified at the hearings, lending support to the phaseout of surface coal mining but declining to endorse any particular proposal. To a question from Ohio Representative John Seiberling about her organizations' position on Hechler's bill, she evasively noted that even H.R. 1000 "has regulatory provisions for the phasing out period." Although Dunlap did not explain it to the subcommittees, in the interim since the last congressional hearings the Environmental Policy Center staff had researched the length of time needed to phase out surface coal mining and came to the conclusion that Hechler's timetable was impractical. The EPC had sent John McCormick to talk to manufacturers of underground mining equipment and they had told him it would take seven years to build the machinery necessary to open enough deep mines to replace phased-out strip mines. Based on this information—which was certainly questionable—

the Environmental Policy Center staff believed they would have to work for a regulatory bill. A seven-year phaseout was unlikely to pass, and only by backing a control measure could they ethically argue "that the lights won't go off." This explains why the bulk of Dunlap's hearing testimony consisted of a detailed list of "minimum requirements" that NCASM and EPC expected to see in final legislation. Though far-reaching and comprehensive, the requirements were much less controversial than outlawing stripping. One suggested provision called for operators performing surface mining on slopes greater than 14 degrees to ensure "that no overburden or debris be placed over the bench unless the applicant can affirmatively demonstrate that the proposed mining method will effectively prevent sedimentation, landslides, erosion, or acid, toxic or mineralized water pollution and that such areas can be reclaimed as required by the Act." Activists had pointed out that steep-slope strip mining could not be done without frequent slides and polluted water, for reasons that had to do with the landscape as well as the ineptitude of agencies given the task of enforcing such rules, but the provision implicitly discounted their experience.[10]

Along with NCASM and EPC, the Sierra Club expressed a wavering position on federal legislative action at the spring hearings too. Citing the "absence of adequate land use planning, a coordinated energy policy, stringent performance standards, strong enforcement, and more advanced technology," the conservation group sent an official policy statement advocating a gradual phaseout of stripping, at least on slopes greater than 20 degrees and in the arid regions of the West. In his testimony, however, Peter Borelli seemed to suggest that the position of his organization was not so far away from the position of the coal industry. "[W]e wholeheartedly agree that a balance, of course, is necessary if our economy and environment is to coexist," he said, "and I know of no other social arrangement that has worked in this country." Like Dunlap, Borelli presented a list of expectations of a regulatory bill. In the interim between passage of legislation and eventual phaseout, or in the more likely case that Congress only regulated stripping, he advised, the Sierra Club hoped for the establishment of an abandoned mine reclamation fund (sustained by a severance tax), EPA oversight of any federal regulatory program (or at least a requirement of the agency's concurrence in decisions made by the Department of Interior), broadened citizen participation in enforcement, and prohibition in selected areas. Shortly after making these remarks, Borelli wrote to Representative Saylor that while the Sierra Club looked favorably on H.R. 1000, other proposals, certain regulatory proposals, deserved "serious consideration."[11]

Other long-established conservation organizations also took ambiguous positions on strip mining controls, making statements that could be read as support for a partial ban as well as a readiness to accept only regulation. Representative John Franson communicated the National Audubon Society's support for prohibition where damage to the land and water would be irreparable, but added that this determination should be made following provisions set out in a good reclamation law. Izaak Walton League environmental associate Nancy Matisoff relayed a policy statement from her organization in support of prohibition of stripping on slopes greater than 14 degrees, "unless an operator can positively demonstrate that sedimentation, acid run, landslides, and other damage can be prevented." In addition to this qualified support for a ban, the IWL's position centered around "a strong federal-state regulatory apparatus," including federal review of state permit and enforcement programs.[12]

By the end of the spring hearings, with support for outlawing surface coal mining obviously weak among national environmental and conservation organizations, previously supportive members of Congress became reluctant to endorse any sort of abolition bill. When Ken Hechler's assistant Ned Helmes was contacted by one of Senator Ted Kennedy's legislative aides about supporting H.R. 1000, he was told that "they were backing off because [Senator] Nelson's office had told him Nelson was backing off." The reason, Helmes told Hechler in a memo, was that "they felt the environmentalists weren't seriously pushing the abolition position anymore." Helmes tried to convince the aide that prospects for passing an abolition bill were better in the House than in the Senate, and he explained Dunlap's "careful" testimony before the Senate committee as moderation necessary to avoid the wrath of Senator McClure and other industry supporters. But the aide ended the conversation with only a promise to "let us know when they made a final decision."[13]

Passage of H.R. 11500 and the First Presidential Veto, 1974

In September 1973, the Senate Committee on Interior and Insular Affairs reported out S. 425, a strong control bill that suggested committee members had been persuaded by at least some of the arguments of strip mining opponents, particularly their concern about erosion and landslides. The proposed act established a federal-state regulatory framework, with a federal Office of Surface Mining, Reclamation and Enforcement (OSM) that would conduct its own inspections and oversee state programs. To guide these state programs, the legislation set minimum standards for permitting and bonding procedures as well as concurrent and postmining reclamation. Among these standards, applications for permits had to include the written consent of surface owners where surface and mineral rights had been separated. The bill also prohibited stripping within the boundaries of national parks, the National Wilderness Preservation System, and the Wild and Scenic Rivers System, and it left open the possibility for declaring other areas unsuitable for surface mining. In a nod to the concerns of organized labor, contracts awarded for the reclamation of abandoned mined lands were to include provisions giving an employment preference to individuals whose jobs had been adversely affected by the control legislation.[14]

With the expectation that the House Committee on Interior and Insular Affairs would report out a regulatory bill in February, and that the full House and Senate would take up consideration of the respective bills sent to them by spring, activists organized a weekend meeting of the national Coalition Against Strip Mining in Washington, D.C., for late January. Prior to the gathering, NCASM leaders sent out a memo to members explaining that they wanted to bring folks together "to work out strategies to shape public opinion in favor of a phase-out and to reach agreements on legislative priorities." It is not clear what exactly happened during the weekend—there are no existent minutes or reports—but key leaders among the 125 participants returned from the meeting determined to work for a good control bill while still calling for a phaseout. This included Reverend Lloyd and Linda Johnson, who informed Appalachian Coalition members that they wanted gradual abolition but also hoped the main regulatory proposal, H.R. 11500 would be strengthened, "as it has some very weak spots." A letter-writing campaign was in the works, they said, and other memos with "helpful factual info" would soon follow from the Environmental Policy Center. According to legislative aide Ned Helmes, the EPC was now in open disagreement with Hechler's phase-out timetable and "supported the Udall bill [H.R. 11500] with four strengthening amendments."[15]

In summer 1974, after the Committee on Interior and Insular Affairs reported out H.R. 11500 and a vote by the full House neared, organized labor announced its support for regulatory legislation too. United Auto Workers legislative director Jack Beidler called on House members to support the control bill, along with unspecified "strengthening amendments." While acknowledging his union's interests in "improving the energy situation" in the United States, Oil, Chemical, and Atomic Workers president A. F. Gospiron insisted "that strong environmental standards must be observed if we are to protect our land and its use and the citizens' quality of life." H.R. 11500 would allow for increased production of coal, he said, while addressing the environmental problems associated with strip mining and land reclamation. The UMW also came out in favor of the bill, but not until after a contentious meeting of the International Executive Board, where the debate was not over the merits of controls versus a phaseout, but rather the need to pass federal regulation at all. Once the exact wording of the House measure became available, President Arnold Miller called the Board into a special session to consider an endorsement. Its members heard presentations from both sides of the issue, first from Representative Patsy Mink, chair of the House Subcommittee on Mines and Mining, who spoke in favor of H.R. 11500. She was followed by William M. Kelce, a former Peabody Coal Company official and head of the Alabama Surface Mining Reclamation Council, who spoke against the bill. After a heated debate, Miller cast a tie-breaking vote and the union went on record in favor of H.R. 11500. "Everyone accepted the need to protect the jobs of our members," Miller later explained, "while protecting areas where our members and their families live." Mink had assured the Board that the legislation would not cause job loss but would likely create new jobs because reclamation standards would encourage surface mining techniques, such as the modified block-cut method, that required more workers.[16]

Despite the support for H.R. 11500 among environmentalists and organized labor, however, at least a few organizations and individuals still pressed for a more radical solution to the problems facing residents in the strip coalfields of Appalachia. In mid-February, the board of representatives of the Council of Southern Mountains adopted an unambiguous statement in favor of abolition. Their first demand was that strip mining be stopped, to protect the mountain landscape and to prevent the disappearance of more deep mining jobs. The board also called for the expropriation of the coal industry if improvements were not made in mining. "If the present managers of coal companies cannot mine coal to benefit everyone," it declared, "then they should turn it over to people who can, like the people who work in the mines." In July, other organizations made public statements in support of a phaseout. Appalachian Coalition co-chair and Save Our Cumberland Mountains leader J. W. Bradley sent a telegram to West Virginia representative Ken Hechler, and presumably other members of Congress, declaring that SOCM "wholeheartedly supports . . . H.R. 15000," Hechler's abolition bill. Wilburn C. Campbell, bishop of the Dioceses of West Virginia and president of the Appalachian Peoples' Service Organization, also sent a telegram in support of H.R. 15000, declaring that "strip mining must be stopped, [to] end the incredible hardship and abuses to people, and to end the devastation of our land for profit. There can be no compromise with this obscene evil."[17]

Through the spring and summer, Representative Hechler received many other letters and telegrams in support of abolition, most but not all from West Virginia and good portion of them from individuals unaffiliated with any particular organization. Among the correspondents, Charles E. Crank Jr. of Huntington described himself as "a concerned citizen and a Christian" who wanted to see strip mining replaced with deep mining, with health and safety standards enforced. "I do not believe the big companies need to make the kind of profits they do," Crank explained, "I believe industry has an obligation to society." Writing from Celina, Ohio, Jean Jones forwarded petitions signed by thirty local residents, addressed to Congressman Tennyson Guyer, in support of H.R. 15000. West Virginia University student Patrick Neal reported that the West Virginia Democratic Youth Conference had passed a resolution calling for "immediate and complete abolition of strip mining." And Alice Slone, director of the Lotts Creek Community School in Hazard, Kentucky, praised Hechler as "the Representative with vision!" She told him to keep up the good work and to either immediately stop or phase out coal surface mining.[18]

Bolstered by public support, Congressman Hechler made numerous public statements against regulatory legislation and in favor of outlawing surface coal mining in 1974. In May, he called H.R. 11500 "a woefully weak compromise, filled with the kind of loopholes the coal and utility lobbyists have written into state strip-mine regulation for the past 30 years." After returning from a trip to the West Virginia coalfields over the Fourth of July holiday, Hechler reported a willingness among some of his constituents to use direct action to stop strip mining. People had told him "that unless Congress stops temporizing with strip mining, they're going to take matters into their own hands just as the patriots of two centuries ago did at the Boston Tea Party and other memorable revolutionary acts." Rather than fiddle with a regulatory bill, the West Virginia representative argued, it would be "better to go all-out for H.R. 15000." In mid-July, he offered his abolition proposal as a substitute for the committee legislation, but the amendment effort failed on a vote of 69 to 336.[19]

The regulatory bill that was at the center of so much concern did include language that would make committed opponents like Hechler cringe. It contained vague terminology dealing with water quality and erosion, which obliged operators to "minimize the disturbance to the hydrologic balance at the mine site" and "control as effectively as possible erosion and attendant air and water pollution." Perhaps even more ominous, the act established a procedure for "the regulatory authority" to grant exceptions to the environmental standards to permit applicants or permit holders "if the regulatory authority issues a written finding that one or more such standards cannot reasonably be met." But H.R. 11500 also contained a number of provisions to appease the opposition, and the report accompanying the bill explicitly acknowledged the growing public concern that had compelled Congress to act. Across the nation, it said, church organizations, environmental and public interest groups, and others have reacted to the "excesses" of surface coal mining by demanding federal legislation banning the practice. Because the committee was concerned about energy requirements, members did not feel they could or should go to this length to resolve the problems. Yet the committee was "equally convinced that equity requires that environmental and social costs which have heretofore been relegated to off-site property owners and to the community at large, must be borne by the producers and users of coal." This would be accomplished, the report explained, through a system of federal standards, federal funding, and the expertise of a federal Office of Surface Mining, which together would "greatly increase the effectiveness of State enforcement programs operating under the Act." As in the Senate bill, the House measure parceled out authority to the states and Department of Interior (rather than the EPA), included a provision for designating lands unsuitable for mining, and required written consent of surface owners in cases where mineral and surface rights had been separated.[20]

In December 1974, after the Senate passed S. 425 and the House passed H.R. 11500, both measures moved to a conference committee. During and after the markup process, it became even clearer where national environmental groups stood on strip mine controls. Before the conference committee met, the national Coalition Against Strip Mining sent a telegram to members of Congress declaring its support for "a well-planned, orderly phase-out" and urging them "to seriously consider and support the merits of H.R. 15000." Signatories to the telegram included the Environmental Policy Center and Sierra Club. After an agreed bill came out of the conference committee and went back to the House and Senate for final approval, however, Montana representative John Melcher sent a letter to all of his colleagues listing its supporters, which included the EPC and Sierra Club, as well as the Wilderness Society, National Audubon Society, Friends of the Earth, Izaak Walton League, National Farmers' Association, and the National Grange. With such solid backing from major environmental and farm organizations, as well as various labor unions, the agreed bill sailed through the House and Senate.

Because supporters expected a veto, Louise Dunlap led an appeal to vice presidential designate Nelson Rockefeller to intercede with President Ford to change his mind. At the end of the month, however, the president did veto this first of two control bills sent to him.[21]

Second Presidential Veto, 1975

When Ken Hechler first proposed legislation to outlaw surface coal mining in 1971, he told the members of the House Committee on Interior and Insular Affairs that his objective was not simply to prod legislators to pass a strong control bill but to actually ban stripping. For the next few years he remained steadfast in this position and became increasingly frustrated with the groups and individuals using the threat of abolition to get better regulatory legislation. In January 1975, on the eve of another weekend gathering sponsored by the national Coalition Against Strip Mining, Hechler expressed some of this frustration in a letter to activists. Yet in the same missive he also made his first bow to "practical politics." "My people in West Virginia and people throughout the nation," he wrote, "are getting more and more cynical about compromising politicians, Washington environmental groups who settle for the lowest common denominator, and those who enjoy the transient glory of winning a few commas or semi-colons while the people and the land continue to be exploited and destroyed." The leadership of Washington-based environmental groups had repeatedly succumbed "to the temptation to move farther and farther away from abolition, to compromise and weaken their position long before it was strategically necessary, and to enable the coal exploiters to move the whole focus of the debate progressively and inexorably over toward greater freedom to exploit." Activists needed to strengthen the backbones of both the EPC and Sierra Club, Hechler argued, and "DEMAND THAT THE PEOPLE'S WILL IN YOUR AREAS BE TRANSLATED INTO MORE AGGRESSIVE ACTION HERE IN WASHINGTON." In this statement to "Friends of the Coalition Against Strip Mining," however, Hechler showed signs of wavering himself. While he encouraged support for his recently introduced abolition bill, he also noted the importance of defining "the minimum standards you would accept in a regulatory bill." For Hechler, these included a ban on steep slopes (greater than 20 degrees), no mountaintop removal, prohibition of mining in national forests, national grasslands, and alluvial valley floors, a ban on impoundments for coal waste disposal, written consent of surface owners in all cases where surface and mineral rights had been separated, and tough citizen suit provisions.[22]

Despite the ambiguity of Hechler's letter, some of the participants in the NCASM gathering remained dedicated to outlawing surface coal mining. When it became clear to him that "Louise Dunlap was not going to adamantly insist we have a ban on mountain stripping," Pineville, Kentucky, activist Stephen Carl Cawood left the meeting. He went home "and urged those folks here who are still fighting to simply drop their support of her and restrict their efforts to lobbying their congressman individually for a ban." Yet something important had happened to the whole movement when Hechler stated his minimum requirements for a good control bill. At the end of January, NCASM issued a statement declaring its support for "the substance" of Hechler's phase-out proposal and calling for an immediate ban on steep-slope mining, no mountaintop removal, prohibition on mining in alluvial valley floors and arid and semi-arid regions, a ban on stripping in national forests and grasslands, and surface owner consent. But the coalition also stated its intent to push for a strict regulatory bill, and now that Hechler had compromised—ironically in a letter about frustration with past compromises—the congressman and environmental groups were no longer working at cross purposes with one another.[23]

By March 1975, both House and Senate committees had once again reported out control bills for floor debate. These proposals were similar to the ones President Ford had vetoed the year before, with a few notable exceptions, such as requiring surface owner consent only where mineral rights were owned by the federal government. Yet, perhaps sensing that the balance of forces was beginning to shift in their favor, opposition to H.R. 25 and S. 7 by elements of the coal industry was much more vocal and apparent. There was now more to gain than lose by opposing basic regulatory legislation. In April a caravan of coal trucks, organized and sponsored by strip mine operators, traveled from the Appalachian coalfields to Washington, D.C., to lobby members of Congress and send a message to President Ford. Starting from the Wise County, Virginia, fairgrounds, the caravan was joined along the way by other independent coal haulers from West Virginia, eastern Kentucky, and eastern Tennessee. Altogether, about six to seven hundred trucks reached the nation's capital. Drivers expressed their reasons for joining in the protest with hand-painted slogans and banners on their vehicles, such as "Protest H.R. 25—We Don't Need Another 25,000 Unemployed" and "Mr. President Save Our Jobs—Veto Bill 25 & Bill 7." In an interview along the roadside, one trucker claimed that the regulatory legislation "will about ban all strip mining" and have a detrimental impact on local and regional economies. Another admitted that there needed to be a control law but suggested that his home state of Virginia enforce the one it already had. Once the caravan reached Washington, D.C., participants organized a motorcade and pickets in front of the White House and Capitol, and small groups systematically visited members of Congress. A few industry representatives also met with federal energy administrator Frank Zarb, who told them that both he and Ford were sympathetic to their concerns.[24]

Despite coal haulers' lobbying, however, by early May the full House and Senate had passed H.R. 25 and S. 7, and a conference committee reported out an agreed version of the control legislation. Reconciling the two proposals was fairly easy since there were only a few significant differences between them. Where changes were made, the regulatory legislation became less rather than more stringent. The House bill's outright ban on mining alluvial valley floors west of the hundredth meridian, for instance, was dropped in favor of the Senate bill's selective prohibition only when such stripping would have a harmful impact on farming or ranching. The conference committee also deleted S. 7's provision requiring the secretary of interior to continually evaluate losses or shifts in employment as a result of enforcement of the act, and it allowed citizen suits only against regulatory agencies and mine operators. The Ford Administration had expressed concern that S. 425 would allow citizen suits against any person for a violation of the act. The committee did retain language in both bills requiring surface owner consent, but only when the mineral rights separated were federally owned.[25]

Once passed by the House and Senate, the compromise bill went to the president, who promptly vetoed the legislation. In his veto message, Ford explained that the regulatory act would put up to 36,000 Americans out of work, raise utility bills, make the nation more dependent on foreign oil, and reduce coal production in the midst of an energy crisis. In hearings before House subcommittees, Frank Zarb elaborated on these objections. He noted that losses from the bill could not really be quantified since there was no way to predict how certain provisions—such as the authority to designate lands unsuitable for mining—would be enforced. Yet he estimated that between 40 million and 162 million tons of coal production would be lost during the first year of implementation as a result of provisions dealing with steep slope mining, aquifer protection, and stripping on alluvial valley floors. The unemployment this loss would entail, a combination of miners directly affected by the legislation and others in sectors of the economy dependent on mining, would be concentrated in certain areas. Some counties in Appalachia, Zarb claimed, could be devastated by the control legislation. Employment created by reclamation efforts would be negligible, and the long lead times and significant capital outlays required to open or expand underground mines would mean that deep mining could not offset these job losses either.[26]

In response to Ford's veto, Congressman Hechler declared that "King Coal continues to control the White House." The president's attack on the legislation as too strong was outrageous, he said, and Ford's claims about coal production and job loss were "totally ridiculous." There was nothing in the bill that had not already been done before under some of the stronger state laws. Yet Hechler planned to sustain the veto *because* the vetoed act fell short of what was needed. Some of the congressman's constituents and other longtime activists pleaded with him to reconsider. Huntington resident Patricia Longfellow asked Hechler to override the veto to give a voice to those most affected by strip mining. Mart Shepard, writing for the Knott County chapter of Citizens for Economic and Social Justice (CESJ)—which had taken the place of AGSLP—also made an appeal for his support. "We no this bill doesn't provide ever thing [*sic*] that is needed in the way of Reclamation," Shepard wrote, "but I am asking you to vote for the short [step] this bill provides." Others seemed to be more inclined to get justice through violence. Charles Douglas, from Barboursville, West Virginia, had served in the Navy during World War II and later in the Marine Reserves, but now felt "sold out by the very institutions I fought for," presumably because Ford had vetoed the bill. This left no other means except "vigilante action" for controlling strip mining. Government was controlled by and existed to serve "special interest groups," Douglas maintained, and the only way to return control to the people "will be by force." Hechler was not willing to go that far. "Although an abolitionist," he said, "I would have settled for a bill which included some hope for the people of the mountains."[27]

Segments of the coal industry viewed the second presidential veto of regulatory legislation as an indication of the weakness of the opposition to stripping, and they took the opportunity to begin to reverse their position on federal controls. In late September, the American Mining Congress declared that uniform national standards for surface mining and reclamation were "not feasible." The trade association cited familiar concerns with diversity of terrain, climate, and other conditions in mining areas, as well as variations of mining methods as the reasons for coming to this conclusion. "The states are properly the focal point for the regulation of reclamation operations," the "Declaration of Policy" explained, and enforcement of state laws, tailored to meet specific physical conditions was "producing good reclamation results." The AMC and its members pledged to endorse adoption of realistic surface mining regulation at the state level and would not oppose federal surface mining legislation that was "realistically designed to assist the states and the surface mining industry." But legislation such as the bill President Ford had just vetoed would, in fact, create "a virtual prohibition on surface mining through the imposition of unrealistic and unworkable provisions," and the Mining Congress could not support it.[28]

The United Mine Workers also moved toward support for state-level regulation of stripping after Ford's second veto. Echoing the concerns of the AMC, a sizable number of delegates to the forty-seventh constitutional convention argued that a federal law would not recognize the problems and conditions specific to various regions. In contrast with what he had said earlier, President Miller explained, "What works in the hills of West Virginia may not work in the plains of Illinois . . . some reclamation standards that would benefit one area could possibly harm another." Surface mining and reclamation were both important to the economy and ecology of the country, he said, and protecting the environment was vital "not only for ourselves but more for the use of our children and our children's children." But the UMW would cease working for legislation to establish federal standards and enforcement and instead work for reclamation laws on a state-by-state basis. Dissent within the union, on surface mining regulation and a host of other issues, had left Miller no other choice than to change his public position on the issue. That dissent could be heard on the convention floor among delegates and coming from union officers, and it revolved partly around fears of the impact any legislation would have on employment. The co-chair of the committee making the recommendation explained that a state-by-state approach was needed "so that we will not put anyone out of work."[29]

President Ford's veto and the failure of Congress to override it also prompted the Appalachian Coalition to "look again at its goals and the needs of groups in the mountains." In November, coalition leaders attempted to hold a meeting in Abingdon, Virginia, to make such a reassessment, but attendance by representatives of local opposition groups was low and another meeting was called for mid-February. At the second gathering, held at the Highlander Center in eastern Tennessee, attendance was better but the meeting exposed the significant divisions within the opposition movement. One SOCM activist observed that "we all have a long way to go before we can all work closely together in what should be a common cause." In mid-March, there was another meeting in Abingdon of the Coalition's newly formed Advisory Council, which included Reverend Lloyd, Phil Roman of Core Appalachian Ministries, Elmer Rasnick of CESJ, SOCM's John Burris and Karen Kasmuski, as well as Frank Kilgore from Virginia and Millie Waters from

Tennessee. They decided to submit proposals for funding to several sources, search for a staff person, and shift funds from Lloyd's Operation Coal to the Appalachian Coalition. By May, the "new" organization was a functioning entity, with Virginian Don Askins hired to serve as central coordinator at an office in Jenkins, Kentucky.[30]

Passage of the Surface Mining Control and Reclamation Act, 1976–1977

Despite two presidential vetoes, supporters of strip mine regulation in Congress were not yet ready to surrender. In March 1976, the House Committee on Interior and Insular Affairs reported out H.R. 9725. Dissenting committee members claimed that the bill would cause a significant loss of coal production, undermine existing state regulatory programs, and increase federal spending. These were by now somewhat typical criticisms that had little basis in fact. But the dissenting representatives also claimed that the bill met the test for being "of the same substance," with only a "few word differences and not a single deleted section" from the vetoed H.R. 25. The Committee on Rules agreed and tabled the measure in the latter part of the month. The House committee then prepared another bill, H.R. 13950, which it reported out at the end of August. Dissenting committee members declared once again that changes in the proposal were cosmetic, that it should be denied "a rule" by the rules committee, and the bill did not go to a vote by the full House.[31]

Between January and March of the next year the House Subcommittee on Energy and the Environment and the Committee on Interior and Insular Affairs held another round of hearings. Testimony from coal operators and corporate officials at the hearings reflected the shift within the coal industry away from support for federal regulation to support for state-level control programs. Speaking for the Kentucky Independent Coal Producers' Association as well as the Kentucky-Tennessee Coal Operators Association, David Smith declared that "the need for a Federal reclamation law does not exist." The need for such a law had passed, he claimed, since thirty-eight states had surface mine reclamation laws "individually suited to their citizen's [sic] economic, environmental, social, geographical, and ecological needs and objectives." Making a novel argument against the likelihood that deep mines might replace any loss in production at regulated surface mines, Smith also said that underground mines would be hampered by "the stringent permit process and reclamation standards for surface effects of deep mining contained in this reclamation act." Speaking for the AMC, Peabody Coal president

Edwin Phelps relayed the trade group's support for state controls. "All major coal producing States have their own functioning programs that regulate surface mining and required sound reclamation," he said. "The national debate that has raged over this issue has outlived the need for Federal legislation."[32]

The hearings also provided an opportunity to the UMW to declare its retreat from federal controls. In their testimony, union officials emphasized the impact the proposed federal regulatory act would have on jobs, the diverse environmental conditions in various states, as well as the better safety record of surface mining relative to deep mining. Joseph Tate, a surface miner and president of a Virginia local, expressed fear that "steep slope" and "return to original contour" provisions in the new committee proposal, H.R. 2, would expand the ever-growing ranks of the unemployed. In addition to being a hardship in its own right, this job loss would devastate the Welfare and Retirement Fund since over half of the moneys paid into the fund came from surface mine employers. Lloyd Baker, president of UMW District 20 (Alabama), argued that "one law with a rigid set of uniform regulations" could not be "workable throughout the country." State legislators knew the conditions and problems of their own states best, he said, and they had enacted workable surface mining laws with this knowledge. Alabama miners also were opposed to a federal law because it would endanger their jobs. "[I]f H.R. 2 has to be the law of the land," Baker stated, "our sincere hope is that it will be completely rewritten so that it controls surface mining as its title states, but does not prohibit surface mining and our surface mining jobs." When questioned by subcommittee members why the UMW had changed its position, Baker explained that his membership in Alabama had been against other federal bills too. The new position simply reflected the victory of another view that had always been held by many in the union.[33]

Governors from Appalachian states were divided in their positions on the proposed control bill. Pennsylvania's governor Milton J. Shapp supported the measure, citing his state's experience with regulation. Challenging the claims by the coal industry and UMW that controls would cut coal production and result in unemployment, Shapp noted that surface mine production in his state had increased substantially between 1971 and 1976, from 26.8 million tons to 38.9 million tons, and the number of strip mine employees had increased from 5,432 to 7,100 during the same period. West Virginia governor John D. Rockefeller (who had defeated Ken Hechler in a primary campaign in 1976) also spoke in favor of the legislation. His state had recently extended the moratorium on strip mining in twenty-two counties for another two years, though it included an amendment to allow surface operations in the counties when landowner-lessors could prove financial hardship and meet reclamation standards. Virginia governor Mills Goodwin was opposed to the bill, however, and Kentucky governor Julian Carroll testified against the provisions that prohibited dumping spoil on slopes and leaving highwalls after mining.[34]

Testimony from activists indicated that they were still in disagreement about accepting regulation rather than a ban. SOCM's J. W. Bradley described the efforts by the federal government to regulate surface coal mining as "short-sighted, unrealistic, and a waste of time." No regulations could make stripping acceptable, he said, and SOCM was demanding a "regulated phase out." Save Our Mountains (West Virginia) executive director Judy Stephenson detailed the corruption and incompetence prevalent among her state's inspectors, implicitly suggesting that more controls would be ineffective.[35] The testimony by other groups, however, revealed that there was a growing contingent of people willing to compromise within the movement. Frank Kilgore, one of the two representatives of the newly organized Virginia Citizens for Better Reclamation (VCBR), said his organization preferred state controls but that it would accept federal regulation if good state legislation was not forthcoming. VCBR was not opposed to strip mining, according to legal counsel Ed Grandis, since it provided jobs and industry in southwest Virginia. But the organization "would like to see more operator responsibility for citizen's property damages," he told the House subcommittee and committee, "plus a stronger reclamation law and enforcement program."[36]

Reverend Lloyd testified at the hearings for the Appalachian People's Service Organization, and he reiterated his Judeo-Christian objection to surface coal mining. He quoted Psalm 24:1 once again and argued that "God created the world as one, whole, interconnected, limited, fragile entity—held together in a delicate balance of interrelationships of all living things." The special role assigned to human beings in this interdependent order, Lloyd said, "is to be caretakers and to live in harmony with creation's on-going, life-giving process." Surface mining was a moral issue, and environmental destruction and the harm caused to people by stripping was evil. "There is no wise answer to strip mining," he declared, "but to phase it out as quickly as we can." Yet Lloyd's rhetoric was out of place in spring 1977. The abolition movement had dissipated by the mid-1970s and Lloyd had been partly responsible for the dissipation. The very short statement submitted by the Appalachian Coalition, an organization he had co-chaired through the decade, urged a prohibition of mountaintop removal and a phaseout of stripping on slopes greater than fifteen or 20 degrees, but suggested the enactment of "severe constraints" as an alternative. The statement also mentioned that H.R. 2 should retain its provisions for citizen participation—as if the group assumed that the bill would be passed.[37]

The other significant figure leading the compromise campaign was Louise Dunlap, and she appeared before the House subcommittee and committee along with EPC representatives Jack Doyle and John McCormick. Anticipating the victory that would come if she could mollify a few coal industry sympathizers, Dunlap made a point to state the support of citizen and agricultural groups for increased domestic coal production, "as well as finding ways to extract both surface and underground coal more efficiently." She called committee members' attention to citizen groups' advocacy of a phase-out of the mountaintop removal method, but also noted the possibility of modifying the method so spoil would be retained on top of the mountain. At the end of her testimony, those who had followed the evolution of the opposition movement as well as the process of revising federal regulatory bills knew quite well what Dunlap was advising the committee to do.[38]

Following the 1977 hearings, on April 20, President Jimmy Carter gave a nationally televised address on energy in which he discussed greater coal utilization but also admonished Americans not to "plunder our environment." Carter had said earlier that he would sign a surface coal mining control bill, and his administration had sent cabinet members to testify in favor of H.R. 2. Carter's secretary of interior, Cecil Andrus, even claimed that neither the House nor Senate versions of the bills in committees were strong enough for him. Andrus wanted a five-year moratorium on stripping on "prime agricultural land" as well as greater protection for subterranean water courses in the arid West and Appalachian mountains. With these signs of support from the president's administration, the House Interior Committee sent H.R. 2 on its way to the full House at the end of April, and the Senate Energy Committee reported out S.7 at the beginning of May. Both bills passed and went to a conference committee in July, where committee member and Kentucky senator Wendell Ford's amendment removing the ban on highwalls was defeated. The conference chairman, Arizona representative Morris Udall, insisted that the highwall ban was "the very heart of the bill," the one provision that would address the destruction of Appalachia's steep slopes. In exchange, however, Ford won acceptance of an amendment allowing variances on the requirement to restore land to "approximate original contour" as well as fewer restrictions on mountaintop removal, bringing the bill into line with the Senate version. He also got an exemption for small mine operators (those producing less than 100,000 tons per year) from federal environmental standards for eighteen months, a compromise between the House version's nine-month exemption and the Senate's two-year exemption.[39]

On July 12, the conference committee reported out H.R. 2, which was soon enacted as the Surface Mining Control and Reclamation Act. In addition to banning highwalls, allowing variances on the approximate original contour requirement, scarcely regulating mountaintop removal operations, and granting exemptions to small mine operators, the bill established a federal-state partnership of enforcement. States would operate their own control programs as long as they met or surpassed the standards outlined in the act. The Department of Interior was charged with approving and overseeing state programs and it had the power to take over state programs if they failed to enforce SMCRA. But surface coal mining regulation was put largely in state hands. Dumping debris on steep slopes was banned by the bill, a provision bitterly opposed by coal operators and one of the few areas in the legislation that had become more stringent over time (the original House bill had prohibited dumping overburden and spoil on slopes greater than 14 degrees). The act also set up an abandoned mine reclamation program, to restore orphan strip lands, funded by a fee on operators of thirty-five cents per ton of surface-mined coal and fifteen cents per ton on deep-mined coal, with at least 50 percent of the fee returned to the state or Indian reservation in which the coal was mined. Yet the legislation allowed for new mining on alluvial valley floors under limited circumstances as well as stripping without the consent of surface owners if permitted under state law. Surface owners whose land had underlying federal coal deposits had to give their consent before the mineral rights were leased and mined, but the secretary of interior reserved the right to override a surface owner's objection if he found that such leasing was in the national interest.[40]

Several environmental and citizens' groups tried to dissuade President Carter from signing SMCRA into law. Representatives from the Appalachian Coalition, including SOCM and Save Our Mountains, announced their opposition to the bill in a press conference in Washington, D.C. The Appalachian Coalition had worked for passage of a federal regulatory bill, the activists acknowledged, but the present bill was so weakened by compromise that it no longer promised effective control of the coal industry or adequate protection of citizens' rights. A press release listed the provisions (or absent provisions) the Coalition found particularly troublesome: an eighteen-month exemption of small operators; recognition of mountaintop removal as an approved mining technique (rather than a variance technique requiring special approval); language allowing for variance from restoration to approximate original contour; failure to impose slope limitations (or a partial ban on contour mining); and failure to fully protect surface owner rights with a comprehensive consent clause. For these reasons, and because SMCRA failed to embody the congressional objective of prohibiting strip mining where reclamation is not possible, the coalition urged a veto. The press release concluded with a declaration that stripping "as a means of energy production must be phased out and replaced by less socially and environmentally destructive methods."[41]

Disregarding the concerns of the Appalachian Coalition, President Carter did not veto SMCRA. At a brief signing ceremony in early August he told a crowd of three hundred guests that the effort to enact a control bill had been somewhat disappointing and that he would have preferred a stricter law. But the "watered down bill" would enhance "the legitimate and much-needed production of coal," the President said, "and also assuage the fears that the beautiful areas where coal is produced were being destroyed." After noting the long lobbying campaign to pass a federal control law, Carter then leaned across the table and kissed Louise Dunlap. When Ken Hechler warned him that there was serious danger that those administering the act would weaken it, the president pointed out that he had a very good secretary of interior, "and I'll remind him of that point." In comments shortly before and after the ceremony, many other prominent figures in the movement seemed to have at least grudgingly accepted SMCRA. Brushy Ridge, Virginia, resident Ransom Meade said the law was better than no bill at all, with good provisions but "far from what we need." One of the original members of AGSLP, Paul Fisty, worried about state enforcement of the act but seemed fairly positive about it. "I'm glad the bill passed. Something is better than nothing," he explained. "They're supposed to put the slope back and eliminate the highwall. . . . Any federal legislation is better than letting the state have control of it." Jack Doyle, at the Environmental Policy Center, bemoaned the failure to ban mountaintop removal and steep slope mining but praised provisions in SMCRA that allowed for citizen-initiated lawsuits against any regulatory authority for not enforcing the law, which provided an opportunity for citizen "watchdogging" of administrative agencies. "The bill is not the best of all possible bills," he admitted, "but it does provide some handles for citizens and is a beginning."[42]

At the end of August, the Office of Surface Mining was officially established within the Department of Interior, and in the fall Walter Heine won appointment to head the new agency. Heine was formerly the associate deputy secretary for mines and land protection in Pennsylvania and had been the first choice for the job among environmentalists who expressed their preferences. But his appointment was the only smooth part of the law's implementation. Congress stalled on giving operating funds to OSM, which delayed the hiring of management personnel and field inspectors as well as the establishment of staff offices in the coalfields until March of the next year. By the spring of 1978, coal operators had initiated an assault on SMCRA in the courts, mounting over one hundred legal challenges to the legislation in more than fifteen separate law suits, although a federal district judge eventually rejected their challenges on constitutional grounds. Congressional allies of the strip mining industry helped operators broaden the attack on SMCRA by scheduling Senate hearings to investigate the law. The hearings were in response to surface miners' claims that regulations were confusing and burdensome to state regulators, small operators could not stand the increased costs imposed by new regulations, and OSM was exceeding the law. Leading the charge at the hearings were Arkansas senator Dale Bumpers, Kentucky senator Wendell Ford, West Virginia senator Jennings Randolph, and senator John Melcher from Montana. Testifying on the first day, West Virginia governor John D. Rockefeller acknowledged that a federal law was needed but claimed that SMCRA regulations had gone too far.[43]

After this rocky start, the federal-state regulatory framework mandated by SMCRA began to fall into place. The central OSM office, with final authority for enforcing the law, was located in Washington, D.C. There were also (initially) five regional offices—each staffed by a regional director, technical staff, inspectors, and three attorneys—as well as district offices within each region. The regional office for Kentucky, Tennessee, Alabama, and Georgia, was in Knoxville, Tennessee, with district offices in Madisonville and London, Kentucky; Birmingham, Alabama; and Knoxville. The regional office for Virginia, West Virginia, Pennsylvania, and Maryland was in Charleston, West Virginia, and federal officials watched over Ohio, Indiana, Illinois, and Missouri from Indianapolis. During the interim period, while states sought approval for their regulatory programs from the Department of Interior, each of the states continued to regulate strip mining but federal inspectors also made regular inspections, responded to citizen complaints, and had the power to assess penalties for violations of interim standards. In fact, during the first five weeks of enforcement federal inspectors visited one hundred operations and closed down eighteen mines, all in the Appalachian coalfields.[44]

The first federal cease-and-desist order was made by Patrick Angel and Bill Hayes, when they responded to a complaint by Hazel King about the Easton Deaton contour strip mine at Clover Fork, in eastern Kentucky. King met the pair at the end of her swinging bridge wearing camouflage fatigues, hiking boots, and pith helmet, and carrying a walking stick. She led them to the mine, which was on an extremely steep slope and, according to Angel, "looked like the aftermath of Hiroshima." There were piles of spoil everywhere, massive, crumbling highwalls, and deep pits of acid water. The inspectors also found a landslide, 200 feet across and 800 feet downslope, moving into Clover Fork. When they wrote the "Order of Cessation," operator David Grooms was stunned and stood motionless, staring at the piece of paper. "All of his equipment and trucks seemed to simultaneously shut down, leaving only silence," Angel later recalled. "Somewhere a white-throated sparrow whistled its pensive song."[45]

chapter nine

Against the Little Man Like Me

Legalized Destruction in the SMCRA Era

Despite initial aggressive action by some inspectors, strip mining in Appalachia was not brought under control after passage of the Surface Mining Control and Reclamation Act (SMCRA). Prohibition advocates were prescient in their skepticism that a new era of environmentally conscious stripping would dawn with enactment of a federal control bill. Even some of the law's chief proponents now acknowledge that the regulatory legislation has been a great disappointment. There is, however, some debate about why SMCRA has fallen short—whether as a result of poor administration and weak enforcement or as a consequence of inadequacies of the law itself. "Even in the worst-case scenario," Louise Dunlap explained twenty years after SMCRA's passage, "I expected the Act to be enforced better than it's been." But she also noted that the situation is not hopeless. "The Act has proven to be sound," Dunlap argued, "the flaws are the White House and the Interior Department." Recent history suggests that she is both right and wrong in her assessment. Proper enforcement of the act does require the support of the president and secretary of interior, and this has not always been forthcoming. But the past couple of decades have also revealed SMCRA's inherent defects. The law has proven to be too reliant on oversight by citizens (who cannot match coal operators' resources and access to lawmakers), too lenient to prevent or rectify various sorts of environmental destruction, and too complex to be applied consistently and effectively.[1]

This last chapter investigates the ways the people of Appalachia attempted to use SMCRA, despite its flaws, to protect themselves and the land. It examines the organizational histories of two state groups, the eastern Tennessee-based Save Our Cumberland Mountains and Kentuckians for the Commonwealth, and concludes with a section on surface coal mining in the 1990s. The first part of the chapter focuses on the early history of SOCM, when it was one of the citizens groups most committed to outlawing stripping at the state and federal levels, and its transformation after passage of SMCRA, when it concentrated on getting the law enforced and finding other ways to constrain coal operators. This is followed by an examination of the origins of Kentuckians for the Commonwealth and its campaign to outlaw the broad form deed, one of the main issues that gave rise to a militant opposition movement in eastern Kentucky during the 1960s. National environmental groups and their representatives had initially included abolition of the broad form deed in their own list of demands for federal strip mine control legislation. As happened with other important concerns of the southern mountain residents, however, the environmentalists bargained it away and Congress wrote surface owner consent out of SMCRA. This was one of the more significant oversights of the act, but it was also an issue on which locally based activists finally had their way. The last part of the chapter takes up another limitation of the law—the failure to ban mountaintop removal, a mining technique that is becoming increasingly prevalent in the coalfields of eastern Kentucky and southern West Virginia. Activists are still struggling to rein in this aspect of the stripping industry.

Save Our Cumberland Mountains

During the 1960s, Tennessee lagged behind other states in passing even minimal control legislation, a fact that was due at least partly to the slow development of an opposition movement in the state. In 1964, when West Virginia, Pennsylvania, Ohio, and Kentucky already had passed and amended strip mine legislation, Tennessee still had not produced its first law. The Tennessee Valley Authority sent a draft control bill to Governor Frank Clement that year, but his administration was in no hurry to introduce and enact such a measure and there was no action on the proposal. The next year, Senator Hobary Atkins introduced a weak control bill, which delayed inspection and enforcement until after mining operations were completed, set penalties too low to function effectively as deterrents to violations, and made no provision for orphan land reclamation. During hearings on the proposal, TVA board chairman Aubrey Wagner and agency reclamation specialist James Curry expressed support for strengthening amendments, but they were the only speakers in favor of stricter regulatory legislation. No citizens' groups requested time to give testimony for controls or a ban. Facing little in the way of public pressure, the House of Representatives voted to exclude the three coal counties of Rhea, Meigs, and Bledsoe in southeastern Tennessee, and the Senate amended the measure to exclude Marion and Grundy Counties as well. The bill was then referred back to the Calendar Committee and, when it came to the floor again, the Senate rejected it.[2]

Sometime after mid-decade, residents concerned about the use of natural resources in their state formed Tennessee Citizens for Wilderness Planning, which, despite its name, worked on numerous pollution and conservation issues, including surface coal mining. Subsequently, in 1967, Tennessee enacted its first strip mine control law. Like other state regulatory legislation, the law defined permitting procedures, mandated reclamation and revegetation of stripped lands, set bonds and provided for conditions of forfeiture, established a strip mining fund for the reclamation of abandoned mines, and set penalties for violations. It gave the power for administering its provisions to the state's conservation commissioner, who was instructed to draft and adopt rules and regulations necessary for carrying out the law, as well as a board of reclamation review. Bonds were set low—not less than $100 but not more than $200 for each estimated acre covered by the permit—and reclamation standards were minimal. Operators had only to grade spoil banks and perform "other soil preparation measures" to provide "favorable conditions for revegetation." The act also allowed the conservation commissioner to grant a variance on the revegetation requirements if an investigation concluded that the conditions of the soil of the stripped land were unsuitable for planting. Rather than stripping being prohibited in an area where revegetation would be difficult, the law left open the possibility of leaving mined land bare until it could grow trees or grasses.[3]

Not surprisingly, the 1967 act failed to control strip mining abuses and this failure prompted annual attempts to amend the legislation through the late 1960s and early 1970s. During the 1972 legislative session, Governor Winfield Dunn, TCWP, and the recently formed Save Our Cumberland Mountains each proposed new bills of their own. The Dunn bill was worked out during meetings between the conservation commissioner and strip operators as part of an effort to abate growing public concern. "There's nothing in that bill that we have not already agreed to," explained an anonymous operator who attended the meetings. "They wanted our agreement on most all that's there. I don't mean it's all that good, but we can live with it." The measure provided for reclamation of abandoned mines with forfeited bonds and fees, allowed operators to take up to a year after an operation was finished to begin reclamation, prohibited mining on slopes greater than 28 degrees, and established a Board of Review that included two representatives from the stripping industry. The stricter TCWP proposal provided for reclamation of abandoned mines with a coal severance tax, required operators to begin reclamation three months after an operation was initiated, prohibited stripping on slopes greater than 15 degrees, and set up a review board composed entirely of persons unconnected to the stripping industry. But the TCWP bill died in the House Conservation Committee, SOCM's measure banning coal surface mining was never even officially introduced, and that left only the governor's proposal.[4]

During consideration of the Dunn bill the governor met with forty members of Save Our Cumberland Mountains, who had come to the capitol to participate in Senate hearings later that afternoon. The meeting was "polite" but tense. "If we were to blow up the strippers' property we would go to jail," one woman explained to Dunn, "but they are blowing up our property." The governor replied that he was open to passing stronger legislation on blasting and, if necessary, outlawing stripping altogether. Yet when the Senate Conservation Committee amended the Dunn bill, in response to SOCM members' complaints about mud slides, polluted streams, and devastation near their homes, the governor's leaders on the Senate floor raised objections. The amendments would "tear up the strip mining industry in Tennessee," declared Senator Daniel Ochmig, sponsor of the governor's bill in the Senate. When the full House considered the amended bill, the Dunn administration provided enough Republican votes to enable Democrats to reject a ten-cent-per-ton tax on stripped coal to fund reclamation of abandoned mines as well as the prohibition of stripping within 300 feet of roads, streams, lakes, or public property. The Senate then passed this version of the bill, nearly identical to the measure originally proposed by the Dunn administration, and the governor signed it into law.[5]

Despite the new control legislation, conditions in Tennessee's strip coalfields did not improve to any significant extent. Both the 1972 law and TVA contracts set a slope limit of 28 degrees, but this standard evidently was not enough to prevent massive landslides and other forms of erosion. Surface mines also polluted hundreds of miles of area streams with acid runoff, killing aquatic life and destroying the water supply for whole communities. In addition, acid spoils made revegetation difficult if not impossible. "We have planted three times," remarked one TVA official about a Campbell County strip mine, "and nothing has taken hold." Even if stripped land could be reclaimed, there were not enough inspectors in the state to ensure that operators would follow the various provisions of the control law. Most strip coal operations were never inspected and, with performance bonds set so low, many operators simply forfeited the portion pertaining to reclamation and replanting, which could be cheaper than actually doing the restoration work.[6]

Surface coal mining had a harmful impact on local and state economies as well. In the early 1970s, nine companies owned most of the coal-producing land in Campbell, Claiborne, Anderson, Morgan, and Scott Counties, which together produced 77 percent of Tennessee's surface and underground coal. Many of these and other land and coal companies had absentee owners, and most paid little if any tax to the state on their property or coal sales. Tennessee had no severance tax on mineral wealth, equipment that should have been taxed as personal business property failed to appear on county assessor books, and land was under-assessed at between $30 to $40 an acre (compared to farmland assessed at between $100 and $500 an acre). Furthermore, coal stripping contributed little in the way of steady wages. Despite the millions of dollars made by companies from coal, residents of strip mine counties were impoverished compared to their counterparts in other parts of the state. Morgan County residents earned $982 a year on average, well below the state average of $2,038, placing it seventh from the bottom in a ranking of all the state's counties. This level of poverty produced heavy net out-migration in the five strip mine counties, draining communities of some of the most able and potentially productive citizens. Strip mining also caused direct economic harm to residents when blasting cracked the foundations of homes but companies denied responsibility. Even worse, state court rulings interpreted broad form deeds as allowing destruction of the surface without compensation to owners of the surface rights.[7]

In the hills and hollows of eastern Tennessee, residents experienced some of the same environmental and economic problems as their counterparts in eastern Kentucky. Because they shared a common history and many cultural values, they also tended to respond to surface coal mining in similar ways. These commonalities are evident in Tennessee farmer Arden Franklin's comments in the Broadside documentary, *A Mountain Has No Seed*. In the film, Franklin explains that strip miners had come close to his home, and that he had written or spoken to local lawyers, state senators and attorneys general, and even U.S. Senator Edmund Muskie about stopping the mining. They all told him that he was in "a bad situation." Franklin's state representative promised that things would change once the new regulatory law was enacted (in 1972), but strippers were "still breakin' windows, sinkin' people's wells, and just acting like a bunch of outlaws." The laws were made by coal company lobbyists, he claimed, "against the little man like me, who makes a living out of the dirt, farmin' anyway I can." Franklin also believed that stripped land could not be "redeemed," which he understood as restoration to conditions that would allow for crop cultivation. "There ain't a thing in the world that can be raised on [strip mined land]," he declared, "[with] acid in there, gullies big enough to bury a big truck in where its washed out, pools of water where holes were left on hillsides." Strip mining tore up God's creation, Franklin explained, and that made it an "awful, awful sin." The only way to deal with these various problems, he argued, was to prohibit the practice and expand deep mining, which would make more jobs available in addition to stopping the ruin of the land. But legislators and the courts were allied with the coal industry, and that left the "little fella" with nothing but his shotgun for protection. This made Franklin uneasy. Neighbors had approached him about joining a "deer rifle committee," but he declined for fear they would all be put in jail. He still had "faith maybe somebody will wake up, the people will, that they'll get together."[8]

While Franklin was trying to figure out what to do to save his land some people were, in fact, "waking up" and "getting together." In September 1971, twelve residents of the state's five largest coal-producing counties filed a complaint with the state comptroller, charging that owners of coal lands were not paying their share of property taxes and requesting the State Board of Equalization to reform the assessment process. The complainants included "two young miners recently fired for signing UMW union cards, several working men employed in Oak Ridge plants, a former county weight inspector who quit his job in protest against the failure to prosecute overweight coal trucks, a local college student, a community worker, and several local women."[9] Shortly after making their complaint, in 1972, the twelve organized Save Our Cumberland Mountains and elected J. W. Bradley as their president, an office he held for five consecutive years. Bradley had been propelled into the opposition movement in 1970 when he and his wife, Kate, were forced off their land by the Shemco Coal Company, which claimed the mineral rights to their property and got an injunction keeping them out of the area where the company planned to strip mine. According to fellow activist

Bill Allen, Bradley was a charismatic leader "who understood only right and wrong; gray was not in his color box." He also kept tight control of SOCM, "not out of a thirst for power but from an implacable desire to eliminate strip mining." Bradley was assisted in working toward this goal by Johnny Burris and Charles "Boomer" Winfrey, both born and educated in the coalfields of eastern Tennessee, as well as Maureen O'Connell, a former schoolteacher from the Midwest.[10]

Like the Kentuckians in the Appalachian Group to Save the Land and People, many SOCM members understood that surface coal mining was linked to other social problems. The organization started with a concern about tax inequities, for instance, and it never stopped working on that issue. "Our children go to poor schools over bad and dangerous roads," complained J. W. Bradley in 1973, "and the little man's taxes go up and up. These big land companies have been starving us out of our fair share of public revenue and it is time to do something about it." One early SOCM pamphlet declared that out-of-state landowners and a few local strip mine operators were gaining benefits, "while Tennessee pays the cost in the form of a ruined environment, stunted economic development, and higher taxes in the future to 'reclaim the mountains.'" When Bradley, Winfrey, O'Connell, and other SOCM representatives traveled to Washington D.C., to testify before committees and lobby members of Congress, they articulated a similarly complex understanding of surface coal mining. By 1975, however, division had developed within the organization about insisting on abolition. Most members argued that the regulatory legislation approved by Congress was too weak to make any impact in the strip coalfields of the mountains, and as an organization SOCM always demanded at least a ban on steep-slope strip mining. But other members were increasingly willing to accept a control law. "Opinions differ as to what the best thing is to do right now," the *SOCM Sentinel* explained at mid-decade. "We are leaving the question with individual members to decide for themselves whether some Federal bill is better than none, whether there is some chance for a better bill in the future, or whether the next Federal bill might be weaker instead of stronger than the present one."[11]

Between 1975 and 1977, SOCM kept a vigilant watch on federal developments and it was among the small number of groups that asked President Jimmy Carter to veto SMCRA. When the president disregarded activists' appeals and signed the bill into law, SOCM began a campaign to block federal approval of the state's control program. In early March 1978, the group released "A Study of Tennessee Strip Mine Enforcement," which questioned the record of the Department of Conservation's Division of Surface Mining

(DSM) since passage of the 1972 state control law. The study listed a number of failings of the DSM, starting with its policy of granting permits to "wildcat" (or illegal) strippers. Other concerns included inspectors' reported reluctance to push investigations and enforcement in general, due to lack of support from their superiors, and the fact that many had been offered bribes by operators. In addition, DSM had not spent any of the reclamation fund, and the Board of Reclamation Review customarily violated the "sunshine law" by meeting in closed-door sessions. The study concluded by requesting that the federal Office of Surface Mining not certify Tennessee, to enforce SMCRA, and calling for the dismissal of Department of Conservation director B. J. Tucker and assistant director Arthur Hope. In response, the federal OSM placed the state on unofficial probation. "We intend to be there looking over their shoulder," explained Director Walter Heine, "and if we get this same indication that they are grossly deficient, we'll move in." Yet little really changed in the short term. Within three months after the study's release, seven DSM staff members left the agency, citing lack of support from administrators. Eventually, in 1984, the state program did lose its federal approval and Tennessee has not regained what OSM refers to as "primacy" since that time.[12]

Many SOCM members were, in fact, skeptical that the problems of strip mining could or would be adequately addressed by the federal government, and the group never stopped working at the state level. Both before and after passage of SMCRA, Save Our Cumberland Mountains pushed state laws restricting and outlawing contour stripping while, at the same time, members used state water quality standards to assert some control over surface mine operations. SOCM achieved a clear legislative victory in May 1976, when the General Assembly enacted and the governor signed the Surface Rights Act. The law had its origins in a fight by White County farmer James Rackley to keep strip miners off his land. Coal operator Julius Doochin insisted that his ownership of the mineral rights to the property allowed him to use any mining method he chose, including stripping, just as operators in Kentucky had claimed in disputes over broad form deeds. This would have ruined Rackley's pasture field, however, and he filed a lawsuit. The case made its way to the Tennessee Supreme Court, which ruled against Rackley in November 1974, and he turned to Save Our Cumberland Mountains for help. SOCM then drafted legislation that required surface owners' consent before a strip mine permit could be obtained for land where surface and mineral rights had been separated. Johnny Burris began lobbying for support of this bill at the 1975 General Assembly and during the next legislative session it passed, making the state the first in the nation to legislate surface owner rights.[13]

In the summer of 1975, SOCM also began a campaign against AMAX, Inc., then the nation's third-largest coal company, which planned to open an area strip mine on 20,000 acres north of Falls Creek State Park. AMAX had acquired only mineral rights, and three obstacles on the surface stood in its way: State Highway 111, several large streams, and several communities, including Pineland (or Piney), in Van Buren County. For the people of Piney, their first knowledge of the strip mining plans came from company representatives who appeared at their doors, bearing checks in hand for the surface rights and threatening to strip their land whether they accepted the payment or not. In July, Maureen O'Connell and Boomer Winfrey traveled through Marion, Van Buren, and Grundy Counties for three days, exploring the site of the intended operation and exchanging information with local people. In October, the residents of Piney packed the school auditorium to listen to the two SOCM staff members, who showed film of the company's dragline equipment along with interviews of residents from an Illinois community that had been fighting AMAX for three years. SOCM was called back a week later for a meeting with state representative Shelby Rhinehart, and then again for meetings with Rhinehart and company lawyers. In December 1975, residents formed the Concerned Citizens of Piney (CCOP), put together a program that ran on Chattanooga television, and began a concerted effort to reach out to such activist groups as the Sparta chapter of the National Organization for Women. By January, CCOP had decided to fight AMAX with a dual strategy, pressuring state regulatory agencies to deny the company permits and pushing for passage of the Surface Rights Act. Later that year, the Water Quality Board denied AMAX a permit and, in 1977, the board denied the company's appeal, at which point AMAX decided to leave the state. Cumberland Plateau communities that had stood in the way of the planned operation won a huge victory, while SOCM gained large numbers of new members and leaders and spread into neighboring counties of the state's southern coalfields.[14]

Yet high coal prices and the imminent implementation of SMCRA made 1978 and 1979 a period of lawlessness, characterized by large numbers of wildcat operations and violence against activists. At a 1978 water quality hearing in Wartburg (Morgan County), attended by a number of well-known troublesome operators, William Goodman and some of his employees interrupted citizen testimony with catcalls and rude comments. Following the hearing, Goodman walked up to J. W. Bradley and punched him in the face. Maureen O'Connell and Annetta Watson then stepped in between the two men, at which point other strip mine employees came in swinging, hit one of the women in the face, and threw both of them on the ground. SOCM's aggressive stance against illegal mines also may have prompted operators to re-sort to arson. When strippers started mining their land without a permit in 1979, Sam and Roberta Baker complained to state officials and the operation was shut down. The Bakers were already involved in efforts to prevent strip-ping all along nearby Douglas Branch and, when their house burned in Oc-tober, SOCM members assumed it was done by strip operators as a reprisal for their activism. Less than a month later, and two weeks after he spoke out against nearby wildcat operations, Campbell County SOCM member John Johnson returned from an errand to find his home burning. The following year, in July 1980, Millard and Mable Ridenour began receiving threatening phone calls and then their house went up in flames. They too had been in-volved in efforts to stop an illegal strip mine. No one was ever charged in connection with any of the fires, however, so it is impossible to conclude with certainty that surface mine operators were responsible for any or all of them.[15]

By the ten-year anniversary of its founding, SOCM had been involved in many battles. Throughout most of the 1970s the group played an instru-mental role in the struggles to pass state and federal legislation outlawing surface coal mining. After passage of SMCRA and dissolution of the abolition movement, the Tennessee activists turned to enforcement of the law. This work gave staff much expertise in the drafting and enforcement of regula-tions, including their use in stopping strip mining altogether, but it did not usually lead to lasting improvements in the state's regulatory program. In ad-dition, group leaders faced the problem of intermittent and short-lived in-terest in fighting stripping. "After the battle to fight a strip mine permit," Bill Allen explained, "most of the citizens SOCM organized dropped out," draining resources and leaving no structure to sustain involvement in new commu-nities. To address some of these problems, the group held a general mem-bership meeting in July 1981, when they reorganized SOCM's governing struc-ture and set meetings of the entire membership on a biannual schedule.[16]

During the next decade of the group's history, its agenda remained dom-inated by concern about strip mining, but members did not attempt to initiate another movement to outlaw the practice. Instead, SOCM mobilized coalfield residents, filed lawsuits, and lobbied legislators with the hope of improving regulatory standards, achieving better oversight, and shoring up the rights of surface owners. One lawsuit against the Office of Surface Mining for its "re-drafting" regulations remained tied up in the courts, while another demand-ing that the agency collect $40 million in unpaid fines was thrown out. The group did achieve an important victory in 1987, however, when the state leg-islature enacted a law establishing a process for reuniting mineral and surface estates. That summer, SOCM also joined with other groups from across the nation for a "Citizens' Coal Summit," in Lexington, Kentucky. Some of these groups continued to meet after the summit and, two years later, they formed the Citizens Coal Council to coordinate the work of activists in various parts of the country and enhance their representation in Washington, D.C.[17]

Kentuckians for the Commonwealth

One month after the last round of congressional hearings on strip mine control legislation, in the early morning hours of 3 April 1977, heavy rains began to fall in eastern Kentucky and southern West Virginia. Within twenty-four hours, four to seven inches of rain came down on ground already saturated from the snowmelt of a hard winter. The surface runoff made its way out of the hollows into creeks and the tributaries of rivers. The rivers quickly reached flood stage and in some areas overflowed their banks. By April 5, when the downpour finally stopped, between eight and fifteen inches of rain had fallen, and downtown Harlan, Pineville, and Williamsburg, Kentucky, were under water. The flooding was responsible for four deaths, left hundreds homeless, and caused at least $175 million worth of property damage, largely borne by the indigent residents living along the floodplains of the affected rivers. Although the rain was a natural occurrence, the high water that resulted was at least partly a human-made disaster. The silt basins and sediment ponds at strip mine sites in the area were outmoded and generally ineffective. As a consequence, operations where land had been stripped of vegetative cover shed tons of silt and sediment into creeks and branches, the mud clogged local streams and rivers, and this exacerbated flooding. While operators acknowledged that sedimentation had played a role in the floods, however, they shifted the blame to other sources of sediment, such as farms, lumbering, and commercial development. Tom Duncan, president of the Kentucky Coal Association, went so far as to say that wide strip mine benches and mountaintop removal actually reduced runoff and erosion. State Division of Forestry director Elmore Grim suggested otherwise. "There is more disturbance by strip mining," he said, "and it was a definite contributor to the flooding."[18]

On July 26, just days before President Carter signed SMCRA, Congress held hearings on whether or not strip mining caused the spring floods. Speaking for the Appalachian Coalition was Jack Sapdaro, a member of Save Our Mountains and a mining engineer employed by the West Virginia Department of Natural Resources to review all drainage control plans for strip mines in the state. Sapdaro declared that stripping had dramatic effects on the hydrologic characteristics of a watershed. "It has been shown in various studies that peak flow rates during storm periods increase by a factor of five after strip mining has occurred in a watershed," he said. "Peak flow has been positively correlated with the percent of area disturbed during strip mining." As was made evident in April, existing sediment control dams and ponds did not provide protection from flooding and sediment damage, and SMCRA was not going to bring much improvement. "The drainage control system required on all new strip mines [by SMCRA]," Sapdaro argued, "is completely inadequate for storage of runoff generated by rainfall exceeding the one- or two-year frequency storm." The only solution to the many problems caused by coal surface mining was a gradual phaseout. "Strip mining is inherently a destructive and undesirable method of recovery," he concluded, "and should not be permitted unless a dire emergency situation exists."[19]

Back in eastern Kentucky and southern West Virginia, residents displaced by the floodwaters experienced great difficulty finding affordable housing outside the floodplain, a difficulty compounded by the fact that most undeveloped land was owned by land and coal companies. When the federal department of Housing and Urban Development delivered trailers to be used as temporary housing, they stood empty alongside the road because there was no place to put them. The land and coal companies were unwilling to lease or sell land overlying coal they might want to mine in the future and so were not welcoming to prospective occupants. The floods thus produced a growing awareness of the monopolistic land ownership patterns in the region. Combined with the blame local people were already casting on strip mine operators for the disaster, this awareness helped to transform new community groups residents had established to assist flood victims.[20]

New groups had appeared first in the eastern Kentucky counties of Harlan and Martin, which had suffered the greatest damage in the state. In Harlan County, affected residents on Clover Fork organized the Cloverfork Organization to Protect the Environment (COPE), and Cranks Creek residents established the Cranks Creek Survival Center. The two groups wanted to have their watersheds declared unsuitable for mining to prevent more environmental damage and flooding in the future, but they grew discouraged when they realized the number of active operations and the amount of money it would take to fight them. Northeast of Harlan, on the West Virginia border, Martin County residents organized the Flood Preparation Group to respond to community needs after the flood, but they also discovered many obstacles in their way, including inadequate land for resettlement. Eventually, the membership of the Harlan and Martin County groups formed the core of the Kentucky Fair Tax Coalition (KFTC), which became Kentuckians for the Commonwealth (with the same acronym).[21]

In West Virginia, Mingo County residents established the Tug Valley Recovery Center. As happened in eastern Kentucky, however, residents soon realized that there was little land open for relocation in either Mingo or Logan Counties, where flooding had been the most serious. In response, in late April, leaders of the Tug Valley organization called more than fifty other groups and numerous individuals from throughout the state to a meeting in Williamson, West Virginia, where they formed the Appalachian Alliance. The primary objective of the new organization was to "support individuals and communities which are working to gain democratic control over their lives, workplaces, and natural resources." As one of the Alliance's first political acts, in August the coalition joined with other regional groups in calling on Carter to veto SMCRA. The next year, in January, seventy-five people from several states met at the Union College Environmental Center in Barbourville, Kentucky, where they elected a steering committee, formalized membership, and made plans to raise $30,000 for general operating expenses, including the salary of one staff person. Shortly after this meeting, the Alliance established a Task Force on Land Ownership and Taxation to study land ownership patterns in eighty Appalachian counties in Alabama, Kentucky, North Carolina, Tennessee, Virginia, and West Virginia. Funded by the Appalachian Regional Commission, the task force investigated who owned the land and minerals in Appalachia and what those owners paid in property taxes, demonstrating in a graphic way the links between the region's poverty, inequality in land and mineral ownership, and inequitable property tax systems. Like the new Martin and Harlan County community groups, this study played an important role in shaping continued concern with surface coal mining and laid the foundation for the formation of Kentuckians for the Commonwealth.[22]

The Appalachian Coalition had responded to the flooding by sending Sapdaro to the congressional hearings, but by the end of 1977 it was primarily concerned with the implementation of SMCRA. During late spring and summer 1978, the group held workshops, with funding from the National Wildlife Federation, "to inform the general public of citizen rights provided in the law and to stimulate public participation in enforcement." The general idea of the workshops was to give folks an understanding of the many technical details of the law and to help them assist OSM personnel. "This law won't stop stripping," Coalition coordinator Don Askins explained, "but it does give you a chance to stop some of the worse abuses and damage; it's up to you to take advantage of the opportunity." Altogether, 125 people from various parts of the Appalachian region attended one of four workshops in Norton, Virginia; Harlan or Hazard, Kentucky; or in Prestonsburg, Kentucky, where the Appalachian Coalition had its office. "Now it finally sounds better to go by the law," former antipoverty worker and eastern Kentucky native Edith Easterling told the Prestonsburg gathering, "than to stand by the bulldozers like we had to do ten years ago." After more than a decade of struggle, battle-weary activists like Easterling had good reason to hope that SMCRA would prove to be a dependable means for controlling surface coal mining.[23]

By the end of the 1970s the Appalachian Coalition had been fighting strip mining for almost a decade. The network had been led by Donald Askins since 1976 and, when he announced his departure in 1979, members decided to call a reorganization meeting. This meeting was held in August, in Jenkins, Kentucky, and the primary agenda item was a discussion about shifting the coalition's emphasis back to the regional and national level or continuing to address local needs (as Askins had done). After a number of participants pointed out that "no one citizens' organization is dealing specifically with the problems associated with strip mining" in Kentucky, the activists decided that the organization's new coordinator would continue to focus on local efforts in the state. Lexington resident Jan Sutherland had already applied for the coordinator position and, after recommendations from Coalition members at the meeting, she was hired to begin work in mid-September.[24]

Although not concerned exclusively with strip mining, local organizations with an interest in controlling stripping had formed in Kentucky in the aftermath of the floods, and they remained organized throughout the rest of the decade. The Concerned Citizens of Martin County had lately participated in the land ownership survey organized by the Appalachian Alliance, investigating the counties of eastern Kentucky. They found that 76 percent of the land in twelve counties belonged to corporations or individuals from outside the area, or to government agencies. Twenty-five corporate and individual owners owned more than a million acres of land, mineral rights, or both, and most of these were headquartered outside eastern Kentucky. These patterns of land ownership were matched by inequitable taxation. The top twenty-five owners, holding nearly half of the land and minerals in eastern Kentucky, paid only one-fourth of the property taxes in the twelve counties surveyed. The Pocahontas-Kentucky Corporation owned 81,333 acres of mineral rights in Martin County, assessed at more than $7 million, yet its annual property tax was only $76. Not only did the company's strip mining ruin the land and water and exacerbate unemployment, but its meager tax burden also probably explained why Martin County had an inadequate public infrastructure.[25]

The primary purpose of the land ownership study had been to galvanize citizen action, and in December 1981, the Alliance hired Joe Szakos as a field organizer in Kentucky. His job was to work with groups in making connections between the study findings and issues of concern in their communities. In late June 1981, Szakos called a meeting in Berea, attended by Kentucky survey coordinator Joe Childers, Council of Southern Mountains representative Laura Batt, Appalachian Alliance coordinator Bill Horton, Appalachian Coalition coordinator Jan Sutherland, and a number of others. They decided to form a coalition to challenge Kentucky's property tax system and to work on revising the tax rates by passing fair tax legislation. The next month, at another meeting in Hazard—which also included members of Concerned Citizens of Martin County and COPE—the focus of the still-unnamed coalition broadened to include the quality of education and land rights, both of which participants considered to be linked to inequitable property taxation. In August, twenty-six people formally established the Kentucky Fair Tax Coalition and elected Concerned Citizens founder Gladys Maynard as chairperson.[26]

In January of the next year, KFTC met in Hazard again and agreed on an agenda for the upcoming legislative session. They decided to focus on increasing the unmined minerals tax (which had been set at a nominal one-tenth of 1 percent of $100 of value by a 1978 law), raising the coal-severance tax (which had been raised from 4 to 4.5 percent in exchange for setting the unmined minerals tax so low), and changing the law that capped state revenue increases at 4 percent. Once the session started, the coalition concentrated exclusively on the unmined-minerals tax and sponsored a rally on the issue in Frankfort. But the tax reform bill died a quick death in the House Rules Committee at the hand of industry ally and House Speaker Bobby Richardson. Because it would be another two years before the Kentucky General Assembly met again, KFTC then began to pursue an entirely different strategy. In Martin County and elsewhere, as historian Jim Schwab explains, the group began to systematically challenge under-assessments through appeals to county boards of assessment and to the Kentucky Board of Tax Appeals. After noncooperation from the state board, and a class action suit by KFTC, the state Supreme Court ruled that coal should be taxed no differently from other real property, and that the near-exemption of unmined coal was therefore unconstitutional. The Revenue Cabinet was slow in establishing a new assessment program, however, and only after a judge threatened to take it over did the agency begin to do its job. As a result, beginning in the early

1990s, land and coal companies were forced to pay significantly higher taxes. In May 1991, for example, the Pocahontas-Kentucky Company paid $296,000 in taxes, a sharp increase from its earlier payments.[27]

In October 1982, KFTC had transformed itself from a coalition into a membership-based organization with local chapters, and subsequently changed its name to Kentuckians for the Commonwealth. As members waged their decade-long struggle to reform taxation of unmined-minerals, the "new" group also took up the issue of broad form deeds, which strip coal operators had been using for decades as justification to destroy the surface in the exercise of their mineral rights. During the 1984 legislative session, KFTC introduced a bill allowing only methods of mining common at the time the deeds were executed. The measure passed and Governor Martha Layne Collins signed it, but the Natural Resources and Environmental Protection Cabinet (NREP) stalled in implementing the legislation. KFTC brought a test case to court and, in April 1985, Judge Calvin Manis ruled in their favor, upholding the constitutionality of the legislation. In another case, the group successfully sued NREP over its failure to enforce the law. On appeal of this latter case, the Kentucky Supreme Court consolidated both cases and, in July 1987, struck down the broad form deed legislation, declaring it unconstitutional.[28]

Following this setback in the state's highest court, KFTC began a campaign to amend Kentucky's constitution, which first required a two-thirds vote in each chamber of the legislature. Both the House and Senate gave unanimous approval to the initiative—indicating the high level of public support for surface owner rights—and it was set to appear on ballots in November. Voters would be able to approve or reject Amendment Number 2, which required surface owner consent before an operator could get a strip mine permit as well as payment for damages to the surface caused by mining. Adopting the slogan "Save Our Homeplace," KFTC's campaign for the initiative brought the group's name and agenda to citizens throughout the state, forged unity among the membership, and succeeded in winning protection for surface owners in eastern Kentucky. After neighborhood canvassing, visits to the state fair, countless newspaper stories, as well as radio and television advertisements, Amendment Number 2 won 83 percent of the vote. Five years later, in July 1993, the state Supreme Court upheld the broad form deed amendment against a challenge by the Lash Coal Company, which had claimed that it was an unconstitutional taking of private property.[29]

With KFTC's victory in the battle against operators' abuse of broad form deeds, a long chapter in the history of Kentucky surface coal mining came to a close. But the 1980s were not without their problems for strip coalfield residents. This was especially true in eastern Kentucky. In a 1987 retrospective, on the eve of the ten-year anniversary of SMCRA, the *Lexington-Herald* reported that the area was still plagued by thousands of abandoned, unreclaimed strip mines and that many more were abandoned by operators every year. These abandoned sites, as well as active mines, continued to cause slides, which forced people from their homes. Both the slides and the long-term erosion put sediment into creeks and rivers, while acid polluted waterways, and blasting cracked foundations and sent flyrock into yards and homes. "The abuse that produced [SMCRA] is still going on," said Thomas J. Fitzgerald, attorney for the Kentucky Resources Council. Some of the abuses might have actually become worse. One coal industry representative claimed that federal regulations had produced more illegal mining. "I don't think there's any question that all these costs on the legitimate operator have created a tremendous incentive for outlaws," said Tom Duncan, head of the Kentucky Coal Association. There were also many intentional misinterpretations by regulators and much abuse of SMCRA provisions granting various exemptions. But the biggest complaint of environmentalists was that the federal control law and regulations were simply not being enforced due to lack of the necessary inspectors and resources.[30]

Mountaintop Removal

Ten and even twenty years after the passage of SMCRA, residents of the Appalachian strip coalfields were still facing some of the same problems landowners had struggled with decades before. Yet their situation was not entirely the same. The strip mining industry underwent significant transformation in the SMCRA era, with more corporate consolidations, the development of ever-larger and more powerful equipment, and the advent of "mountaintop removal." Although underground as well as traditional contour and area mining continued, during the 1980s and 1990s increasing numbers of operators in eastern Kentucky and southern West Virginia extracted coal by lopping off whole mountaintops. In 1973, the West Virginia Department of Environmental Protection permitted only 300 acres for mountaintop removal. Toward the end of the 1980s this number jumped to more than 4,000 acres, dropped to 960 acres in 1994, and increased sharply thereafter, to more than 12,000 acres in 1997. Throughout all of the 1980s, the state agency issued forty-four permits for mountaintop removal mines covering a total of 9,800 acres, but between 1995 and 1998 alone it permitted thirty-eight mountaintop removal operations covering a total of nearly 27,000 acres. As the acreage of mountaintop removal mines increased, the size of individual mines also grew. By the turn of the century, the average mine permitted by the Department of Environmental Protection was 450 acres, roughly equivalent to 360 football fields. Over time, as operators mine contiguous sites, the total area directly affected by a single operation could be as much as 5,000 acres.[31]

The impact of this new type of mining was and is devastating. Mountaintop operations usually replace at least some of the overburden they move to uncover a coal seam, but they also dump a good portion of it in "valley fills," burying streams and disrupting aquatic life. One recent survey of eastern coal states by the U.S. Fish and Wildlife Service found nearly 900 miles of streams buried in this way, with more than half of those in just five of West Virginia's thirteen coal counties. According to Rick Eades, a hydrologist who once worked in the mining industry, these valley fills are not very safe. "What you see in a lot of these valley fills has no engineering method to it at all," he explains, "It's just dirt and rock being pushed over the side of a hill and filling in vertically several hundred feet." William Maxey, West Virginia's former chief forester, also decries the way mountaintop removal has affected the state's woodlands. As of 1997, 300,000 acres of hardwood forest had been destroyed by mountaintop operations, and the Department of Environmental Protection had done little to ensure reforestation, which is what prompted Maxey to resign. In terms of jobs, mountaintop removal is part of a continuous effort by mining companies to reduce their costs of production by reducing its need for workers, thereby worsening the region's unemployment. West Virginia had 100,000 union coal miners at mid-century, but this has since declined to 19,000 miners, only half of which belong to the United Mine Workers. The decline was caused primarily by mechanization and the shift to strip mining, not a decline in coal production (which has actually increased to record levels). It is also the result, however, of the migration of strip mining westward, particularly to the thick seams and nonunion sites of Wyoming's Powder River Basin.[32]

When SMCRA was passed by Congress in 1977, most of its supporters assumed mountaintop removal would occur only infrequently, in exceptional circumstances. And perhaps that explains why the law's provisions defining reclamation after a mountaintop operation are so vague. Operators are required to reclaim the land so that it "closely resembles the general surface configuration of the land prior to mining," or the "approximate original contour" (AOC). This standard could make mountaintop removal mining completely impractical in places like eastern Kentucky and southern West Virginia, where grades often exceed 20 degrees. But it has not been interpreted that way by coal companies, and regulatory officials contend their hands are tied because nobody agrees what AOC really means. SMCRA does allow some exemptions to the standard, if operators submit detailed plans for development of schools, factories, or public parks before permit approval. Yet regulators have failed to require even that much of operators. According to West Virginia Department of Environmental Protection records, sixty-one of eighty-one active mountaintop removal mines in 1997, accounting for 70 percent of the state's mountaintop removal production, did not get exemptions. Still, operators are not rebuilding the mountains either. Instead, they are turning their sites into pastures and rolling fields. The most popular postmining land uses proposed and approved by the Department of Environmental Protection, based on an investigation of the agency's own records by *Charleston Gazette* reporter Ken Ward, are "fish and wildlife habitat" and "timberland," neither of which is recognized by SMCRA. OSM's own investigation confirmed Ward's findings.[33]

Some of the controversy over mountaintop removal came to a head recently, after Arch Coal applied for a permit to expand an existing operation in Logan County, West Virginia. The expansion would have amounted to 5 square miles of "total extraction." In response, ten coalfield residents joined with the West Virginia Highlands Conservancy to sue the Army Corps of Engineers and the Department of Environmental Protection for overlooking provisions in SMCRA, the Clean Water Act, and other laws. By the spring of 1999, the plaintiffs had settled part of the case out of court and Arch Coal had scaled back its planned expansion, after which the company argued it should not be put under the stricter scrutiny mandated when operations impacted so many acres of a drainage basin. Instead, U.S. district chief judge Charles Haden issued a preliminary injunction to delay the permit until the lawsuit was settled later that summer. Arch Coal responded by laying off thirty miners at the existing mine and threatening to shut down its operation and put hundreds more miners out of work. This prompted United Mine Workers president Cecil Roberts, a native West Virginian, to call two rallies in Charleston. At one of these rallies he told the crowd that they had been "kicked in the teeth again by the environmental community," pointing the blame at coalfield residents themselves rather than the energy conglomerates that own the coal companies and their subsidiaries. "The environmental extremists do not want to listen to our ideas for compromises," he said, "because their goal is simply to shut down the nation's coal industry." Of course, none of the opponents of surface coal mining ever publicly voiced support for the abolition of all coal extraction, though there was at one time a significant movement to outlaw stripping, including mountaintop removal. At the close of the twentieth century, however, that movement seemed a part of the distant past.[34]

Having to Fight the Whole System

As the preceding chapters demonstrate, significant levels of opposition to surface coal mining developed in the years after World War II, primarily in the strip coalfields of Appalachia. Initially, opponents assumed that state regulation could effectively reign in operators, and it was in the immediate postwar years that most states made at least tentative steps toward enacting control laws. By the late 1950s, there was also an increasing number of people who favored outlawing stripping altogether. They had witnessed the failure of the first state laws to prevent environmental degradation and had come to view strip mining as fundamentally bad for local and regional economies. Most of these advocates of prohibition, and many of the proponents of regulation, were common people—small farmers, active and retired deep miners, homemaker wives and mothers. The largest portion of them were also rural dwellers. On the whole, the opposition was not dominated by the middle-class suburbanites who play such a central role in the narrative of environmentalism put forward by Samuel Hays. And opponents did not need those more privileged and far-removed folks to pioneer environmental concern before they themselves became critical of eroding and sliding soil, polluted streams, and vanishing fish and wildlife.

Opposition to strip mining emerged from western Pennsylvania all the way to northern Alabama and evolved over more than three decades. Not surprisingly, efforts to control or stop stripping varied geographically and changed over time. Yet there was some consistency and continuity in the perspectives of opponents. Most critics shared a basic concern with what surface coal mining was doing to their communities, including homesteads, farm- land, water supplies, and employment opportunities. They tended to couple this set of worries with the complaint that large coal companies benefiting a few should not be able to make great profits at the expense of "the public." In addition, many opponents made traditional conservationist arguments. They suggested the need for more rational use of natural resources and voiced concern about the disappearance of fish and game. Others, such as Tennessee farmer Arden Franklin, thought surface coal mining was an "awful, awful sin," a violation of God's command for people to be stewards of the earth. Linked to this Judeo-Christian criticism of the industry were critics' claims that stripping destroyed the beauty of the land.

On a basic level, the campaigns to regulate or ban surface coal mining were responses to the detrimental impact strip mining had on communities and homes, regional employment, and the ecological integrity and beauty of local landscapes. The above listing of concerns indicates, however, that pro- ponents of control legislation and abolitionists saw the harm that stripping was doing through a variety of cultural lenses. Traditions and values shaped their understanding of why surface coal mining was wrong as well as the pa- rameters for how to respond to correct abuses and prevent further damage. The most important cultural lens refracting abolitionist arguments, and sig- nificantly influential in shaping grassroots demands for regulatory laws, was an American tradition of veneration for small private property. In letters, handbills, congressional testimony, and other private and public documents, opponents of strip mining often contested the practice as harmful to their own proprietary rights. Talking about his property on Big Fork and Clear Fork of Lotts Creek, in Hazard, Kentucky, Clarence Williams articulated this type of argument in a 1965 letter to the editor of the *Hazard Herald*:

About two years ago . . . [I] sold right of way to [a] strip mining company. . . . Now it makes me sick to look at my land. A nice orchard of more than 25 trees is no where to be seen. Thousands of dollars worth of timber is under the rocks and trash pushed down by machinery. On tops of the mountains all you can see are acid wastes and stagnant pools. . . . With the heavy rains beating down these bare hillsides what will hold the soil? . . . The poor people in the valleys will be covered with the filth leaving no farm land to grow crops. . . . Now the coal operators are trying to take the rest of my farm. . . . Do people think I'm going to sit and let this happen? I don't intend to sit idly by and see the resting place of my ancestors covered by spoils from the strip mine destruction. This is my home. . . . They yell out to save jobs and industry. They don't realize their jobs mean destroying what the other people own.[1]

By the end of the 1970s, a few activists had begun to discuss and work toward democratic control of the natural resources of Appalachia. They sensed that the root of many problems in Appalachia actually hinged on the private control of those various resources. For the most part, however, local and regional movements to control or abolish strip mining never posed a serious challenge to the institution of private property. Instead, they embraced it as the foundation for their opposition.

Traditions and values also shaped the way proponents of a ban organized and struggled against surface coal mining. Remnants of a classical republican ideology compelled small freeholders to defend their land as the foundation for independence, self-reliance, and a virtuous social order. Similarly, a Lockean doctrine of natural rights deemed rebellion a right and duty when any of the other God-given rights were threatened by corrupt, tyrannical rule. The Lockean legacy is particularly noticeable in Warren Wright's opening remarks at the 1971 "Peoples' Hearing on Strip Mining" in Wise, Virginia. "I do not want to disturb anyone here," he said, "but those of you meeting here in harmony with the democratic purpose of this hearing have evinced a degree of the same disrespect and cynicism that some landowners display in violent refusal to defend their property. For we are here to consider and evaluate our position as under the power of industrial corruption; a people's hearing can say but one thing—'Our judges and our legislators, seen in the mass, are too crooked and too heartless for us to trust. We must attempt to bypass them.'"[2]

Facing off strip operators with a shotgun in hand, dynamiting mining equipment, blocking coal trucks and bulldozers, and merely threatening to do any one of these things all gained legitimacy by traditions much older than strip mining itself. Outside observers often acknowledged as much, and they sometimes cited direct action as a legitimate means to defend private property. This is the suggested subtext in a 1967 editorial that appeared in the Louisville *Courier Journal:*

Once again Kentucky is seeing people take up arms to protect their homes and land because the law can't or won't do it for them. . . . There is something that violates the very character and concept of America in the fact that a company in search for coal may enter another man's property against his wishes [by a broad form deed], destroy his land, wreck his livelihood and endanger his home and family and personal safety . . . this flies in the face of the American belief in fair play, the sanctity of private ownership, and the right of every man to have, hold, develop and defend what is rightfully his . . . all of us lose when the rights of any man are crushed, whether by court edict or dictator's rifles.[3]

Even some strip mine employees understood the protests against stripping this way. When Dan Gibson and members of the Appalachian Group to Save the Land and People protected the land of Gibson's stepson with shotguns, a worker at the site supposedly told Dan that he was doing the right thing because "they" had no right to tear up people's land.[4]

Despite the ways in which defense of private property resonated with various observers, however, advocates of outlawing surface coal mining were never successful in achieving their objective. The explanation for this failure is threaded with cruel irony. The insurgent character of the abolition movement to outlaw strip mining was, in fact, the primary reason that governors, state and federal lawmakers, the coal industry, and the United Mine Workers came to accept stricter regulation of stripping. Control bills and laws were concessions offered by legislators and members of Congress to quell the rebellion in the strip coalfields and prevent the opposition from completely abolishing strip mining.[5] But it is also clear that some activists' intentional use of the threat of prohibition as a foil to get stronger regulation was ultimately counterproductive. Once lawmakers and coal operators understood that an ever-larger number of regional and national groups were willing to accept federal regulation, proposed regulatory bills became progressively weaker and a legislative context was created in which it was increasingly difficult for committed proponents of abolition to be taken seriously. In the end, the "foil strategy" helped undermine the opposition as a whole and resulted in a weaker rather than a stronger regulatory law.

Part of the responsibility for the opposition's retreat from the demand for abolition lies with national environmental leaders. When the campaign for abolition shifted from the local to regional and national levels, the influence of practical-minded leaders increased, the distance between the leadership and grassroots activists grew, and a space was opened for greater participation in the movement by long-established (and not-so-long-established) national conservation and environmental groups. The rising leaders and national groups played traditional beltway politics, engaging in compromise long before any sort of decline in the movement required it, and lawmakers responded. Negotiations were a product of the militancy in the coalfields, but the movement's leaders either failed to recognize or willingly ignored this fact, while lawmakers took advantage of the opportunity to address the issue with "reasonable" people. During hearings on regulatory bills, members of Congress seemed downright giddy in exchanges with lobbyist Louise Dunlap—although it is difficult to determine how much of this was due to her willingness to be co-opted and take the movement off course and how much of it was due to the greater ease they felt with a woman who dressed and acted the way they thought a woman (if she was going to be involved in politics) should dress and act. Perhaps all this is also part of the explanation for that brief but very meaningful kiss President Carter gave Dunlap in the Rose Garden, when he reached across the table toward her after signing SMCRA into law. He probably would not have done the same to militant Appalachian activists Eula Hall or Bessie Smith—or they would not have let him.

Some activists seem to have realized the fix they were getting into early on in the abolition campaign, but the way out was difficult to fathom and daunting in its implications. Mary Beth Bingman suggests as much in her recollection of the 1972 strip mine occupation in Knott County, Kentucky. "After [the occupation] I felt more consciously that you can't try to fight on an issue like this without having to fight the whole system," she recalled, "and you can't successfully organize the community to fight such an issue without trying to change the whole system." The Appalachian Group to Save the Land and People had developed out of a sense that strip operators' degradation of the environment was linked to local and regional economic decline. By the early 1970s, Bingman and others were also beginning to understand that stripping was only one part of a larger, unjust system and, more importantly, that the part could not be changed without transforming the whole. That new consciousness was shared by only a few of the opponents of surface mining, however, and there were many overwhelming forces working against the formation of a more comprehensive movement.[6]

notes

Introduction

1. T. N. Bethell, "Hot Time Ahead," *Mountain Life and Work* (April 1969), reprinted in *Appalachia in the Sixties: Decade of Reawakening*, ed. David S. Walls and John B. Stephenson (Lexington: University Press of Kentucky, 1972), 116–19.

2. Samuel P. Hays, *Beauty, Health and Permanence: Environmental Politics in the United States, 1955–1985* (Cambridge: Cambridge University Press, 1987), 144–46; John Opie, *Nature's Nation: An Environmental History of the United States* (New York: Harcourt, Brace and Company, 1998), 350–53; Duane A. Smith, *Mining in America: The Industry and the Environment, 1800–1980* (Lawrence: University Press of Kansas, 1987); Richard H. K. Vietor, *Environmental Politics and the Coal Coalition* (College Station: Texas A&M University Press, 1980). For relevant Appalachian studies historiography, see, for instance, David E. Whisnant, *Modernizing the Mountaineer: People, Power, and Planning in Appalachia* (1980; reprint, Knoxville: University of Tennessee Press, 1994).

3. The still widely accepted interpretation by Samuel Hays argues that environmentalism developed in the years after World War II as a result of new interest in quality-of-life issues, expanding material affluence, increased leisure time, and rising levels of education, all of which brought new values into politics. These concerns and values were rooted in middle-class suburbia and, Hays claims, to the extent others also demonstrated an environmental sensibility, the environmental concerns "worked their way from the middle levels of society outward." Hays, *Beauty, Health and Permanence*, 13.

4. "The sanctity of private property, the right of the individual to dispose of and invest it, the value of opportunity, and the natural evolution of self-interest and self-assertion, within broad legal limits, into a beneficent social order," wrote Richard Hofstadter, "have been the staple tenets of the central faith in American political ideologies." Richard Hofstadter, *The American Political Tradition and the Men Who Made It* (New York: Knopf, 1948). See also Louis Hartz, *The Liberal Tradition in America* (New York: Harcourt, Brace and Company, 1955).

5. In an article on American exceptionalism, for example, Eric Foner suggested that it was "not the absence of non-liberal ideas, but the persistence of a radical vision resting on small property [that] inhibited the rise of socialist ideologies." Eric Foner, "Why Is There No Socialism in the United States?" *History Workshop* 17 (1984): 57–80.

See also Steven Hahn, *The Roots of Southern Populism: Yeoman Farmers and the Transformation of the Georgia Upcountry, 1850–1890* (New York: Oxford University Press, 1983), and Nick Salvatore, *Eugene V. Debs: Citizen and Socialist* (Urbana: University of Illinois Press, 1982).

Chapter One

1. French artist Jacques Le Moyne first designated the mountain region "Montes Apalatchi" while traveling with a Huguenot expedition to Florida in 1564. By the middle of the eighteenth century, however, explorers, mapmakers, and others more often referred to the area as "Alleghenia" or "Allegheny Mountains." In the early twentieth century, physical geographers used "Appalachia" to designate the larger regional mountain system, reserving Allegheny for a smaller subregion within that system, and the older name became the principal geographic term for the area once again. See Donald Edward Davis, *Where There Are Mountains: An Environmental History of the Southern Appalachians* (Athens: University of Georgia Press, 2000), 3–8.

2. Karl B. Raitz, Richard Ulak, and Thomas R. Leinbach, *Appalachia: A Regional Geography: Land, People, and Development* (Boulder: Westview Press, 1984), 11, 14.

3. John Rodgers, *The Tectonics of the Appalachians* (New York: Wiley, 1970), 4, 7, 19–22.

4. Ibid., 5. Anthracite was once mined from 484 square miles in ten counties in northeastern Pennsylvania. The bituminous coal basin of Appalachia, which still has reserves supporting active operations, underlies 72,000 square miles in parts of nine states. Raitz, Ulak, and Leinbach, *Appalachia*, 79–82.

5. Rodgers, *The Tectonics of the Appalachians*, 5–6.

6. J. Marvin Weller, "Cyclical Sedimentation of the Pennsylvanian Period and Its Significance," *Journal of Geology* 38 (February–March 1930): 97–135; Harold R. Wanless and Francis P. Shepard, "Sea Level and Climatic Changes Related to Late Paleozoic Cycles," *Bulletin of the Geological Society of America* 47 (1936): 1178–1206.

7. George DeV. Klein and Jennifer B. Kupperman, "Pennsylvanian Cyclothems: Methods of Distinguishing Tectonically Induced Changes in Sea Levels from Climatically Induced Changes," *Geological Society of America Bulletin* 104 (February 1992): 166–75; George DeV. Klein and Debra A. Willard, "Origin of the Pennsylvanian Coal-Bearing Cylcothems of North America," *Geology* 17 (February 1989): 152–55.

8. Still another way of understanding the origins of cyclothems, particularly addressed to those in the Appalachian Plateau, points to river mouth depositional environments as an important explanation for succession. This hypothesis was presented by John C. Ferm in 1974 and has a number of advocates. See John C. Ferm, "Pennsylvanian Cyclothems of the Appalachian Plateau, A Retrospective View," in *Carboniferous Depositional Environments in the Appalachian Region*, ed. John C. Ferm and John C. Horne (Columbia: Department of Geology, University of South Carolina, 1979), 287–89.

9. Raitz, Ulak, and Leinbach, *Appalachia*, 68–70.

10. Richard White, *The Middle Ground: Indians, Empires, and Republics in the Great Lakes Region, 1650–1815* (Cambridge: Cambridge University Press, 1991); Wilma Dun-

away, "Speculators and Settler Capitalists: Unthinking the Mythology About Appalachian Landholding, 1790–1860," *Appalachia in the Making: The Mountain South in the Nineteenth Century*, ed. Mary Beth Pudup et al. (Chapel Hill: University of North Carolina Press, 1995), 51.

11. Dunaway, "Speculators and Settler Capitalists," 52–54, 61.

12. Ronald Eller claims that the nineteenth-century Appalachian social order actually replicated the Jeffersonian ideal. See Ronald Eller, *Miners, Millhands, and Mountaineers: Industrialization of the Appalachian South, 1880–1930* (Knoxville: University of Tennessee Press, 1982), 3, 16, 43.

13. Henry Shapiro, *Appalachia on Our Mind: The Southern Mountains and Mountaineers in the American Consciousness, 1870–1920* (Chapel Hill: University of North Carolina Press, 1978), ix; Allen Batteau, *The Invention of Appalachia* (Tucson: University of Arizona Press, 1990), 31 and 33.

14. Shapiro, *Appalachia on Our Mind*, 4, 5; Eller, *Miners, Millhands, and Mountaineers*, xvi; Batteau, *The Invention of Appalachia*, 76.

15. Shapiro, *Appalachia on Our Mind*, 60.

16. Edward L. Ayers, *The Promise of the New South: Life after Reconstruction* (New York: Oxford University Press, 1992), 9, 20, 117.

17. Crandall A. Shifflet, *Coal Towns: Life, Work, and Culture in Company Towns of Southern Appalachia, 1880–1960* (Knoxville: University of Tennessee Press, 1991), 13–14; Ayers, *Promise of the New South*, 189–90.

18. Eller, *Miners, Millhands, and Mountaineers*, 56; see also John Gaventa, *Power and Powerlessness: Quiescence and Rebellion in an Appalachian Valley* (Urbana: University of Illinois Press, 1986).

19. Ayers, *Promise of the New South*, 119; David Alan Corbin, *Life, Work, and Rebellion in the Coal Fields: The Southern West Virginia Miners, 1880–1922* (Urbana: University of Illinois Press, 1981), 5; Eller, *Miners, Millhands, and Mountaineers*, 128.

20. Paul Salstrom, "Newer Appalachia As One of America's Last Frontiers," in *Appalachia in the Making*, ed. Mary Beth Pudup et al., 92; Corbin, *Life, Work, and Rebellion in the Coal Fields*, 8; Ayers, *Promise of the New South*, 123.

21. Helen Lewis was one of the first scholars to use a "colonial" model to understand the poverty of Appalachia as a consequence of economic development. See her essay, with Edward E. Knipe, "The Colonialism Model: The Appalachian Case," in *Colonialism in Modern America: The Appalachian Case*, ed. Helen Matthews Lewis, Linda Johnson, and Donald Askins (Boone, N.C.: The Appalachian Consortium Press, 1978). See also Eller, *Miners, Millhands, and Mountaineers*, xxiv–xxv, 85. More recently, historians and sociologists studying Appalachia have applied the core/ periphery model of Immanuel Wallerstein. See Wilma Dunaway, *The First American Frontier: Transition to Capitalism in Southern Appalachia, 1700–1860* (Chapel Hill: University of North Carolina Press, 1996) and Davis, *Where There Are Mountains.*

22. Howard N. Evanson, *The First Century and a Quarter of American Coal Industry* (Pittsburgh: n.p., 1942), 29, 41, 156. Richard Smith quoting Dr. Schoepf in *Proceedings of the Symposium on Strip Mine Reclamation, June 23–25, 1965*, 1, Box 24, Folder 1, "Kentucky Conservation Council and Ad Hoc Committee on Strip-Mining, Jan. 3, 1969–Dec. 31, 1969," Harry Caudill Manuscript Collection, Special Collections,

University of Kentucky (hereafter cited as Caudill Papers). J. A. Bownocker and Ethel S. Dean make scattered references to Ohio farmers using surface mining techniques to extract coal for personal consumption and local sale up to the twentieth century; J. A. Bownocker and Ethel S. Dean, "Analyses of the Coals of Ohio," *Geological Survey of Ohio Bulletin*, 4th ser., 34 (1929): 6, 14, 20, 22–23, 27, 36, 173, 176, 182 , 247, 273–74, 276, 284. See also George Siems, "The Strip Mining of Bituminous Coal in West Virginia: An Analysis of Past and Present Conditions" (Master's thesis, University of Pennsylvania, 1949), 3, and Julian Feldman, "The Development of a Regulatory Policy for the Coal Stripping Industry in Ohio" (Master's thesis, Ohio State University, 1950), 5.

23. Harry Caudill, *My Land Is Dying* (New York: E. P. Dutton, 1973), 57–58.

24. *Coal Age* 25 (May 1924): 197, 797; Robert F. Munn, "The First Fifty Years of Strip Mining in West Virginia, 1916–1965," *West Virginia History* 35 (October 1973): 66.

25. John Hendrickson, "The Development of Strip Coal Mining in Indiana" (Master's thesis, Indiana University, 1952), 41, 69, 78; *Coal Age* 34 (June 1929): 335, 337; Feldman, "The Development of a Regulatory Policy," 5–6.

26. Coal seams were being exploited in Ohio by the early nineteenth century. An advertisement in the Pittsburgh *Gazette*, August 12, 1797, mentioned coal as an inducement for settlement near Steubenville: "Sale of Lots. In the new County Town of Fort Steuben, in the new county called Jefferson, on the bank of the river Ohio. There is a Sawmill close to the town—and the abundance of Pitt Coal, will render fuel a very cheap article forever. Bezabel Wells, 8/1/1797." See Evanson, *First Century*, 265.

27. *Coal Age* 9 (January 1916): 161–62; *Coal Age* 25 (June 1924): 835; W. S. Judy to Governor James Rhodes, 16 November 1964, Box 165, Frank J. Lausche Papers, Ohio Historical Society (hereafter cited as Lausche Papers); U.S. Bureau of Mines, *Minerals Yearbook: Review of 1940*, 787; U.S. Bureau of Mines, *Minerals Yearbook: 1939*, 793; U.S. Bureau of Mines, *Statistical Appendix to Minerals Yearbook, 1932–1933*, 405–07.

28. Carmen Peter DiCiccio, "The Rise and Fall of King Coal: A History of the Bituminous Coal and Coke Industry of Pennsylvania from 1740–1945" (Ph.D. diss., University of Pittsburgh, 1996), 59–60; Munn, "The First Fifty Years," 66–74; Harry Caudill, *Night Comes to the Cumberlands: A Biography of a Depressed Area* (Boston: Little, Brown and Company, 1962), 313.

29. U.S. Bureau of Mines, *Minerals Yearbook: 1939*, 794; U.S. Bureau of Mines, *Minerals Yearbook: 1948*, 311.

30. Kentucky Strip Mining and Reclamation Commission, *Strip Mining in Kentucky*, 8; U.S. Bureau of Mines, *Minerals Yearbook: 1948*, 309; U.S. Bureau of Mines, *Minerals Yearbook: 1955, Fuels*, 66; U.S. Bureau of Mines, *Minerals Yearbook: 1958, Fuels*, 83.

31. Munn, "The First Fifty Years," 67–72.

32. Tennessee Department of Conservation and Commerce and the Tennessee Valley Authority, *Conditions Resulting from Strip Mining for Coal in Tennessee*, 1; U.S. Bureau of Mines, *Minerals Yearbook: Fuels, 1963*, 93, 102; U.S. Bureau of Mines, *Minerals Yearbook: 1958, Fuels*, 48, 81, 86.

33. Caudill, *Night Comes to the Cumberlands*, 313–14; U.S. Bureau of Mines, *Minerals Yearbook: 1958, Fuels*, 87; U.S. Bureau of Mines, *Minerals Yearbook: Fuels, 1963*, 105.

34. Kentucky Strip Mining and Reclamation Commission, *Strip Mining in Ken-

tucky, 14; U.S. Bureau of Mines, *Minerals Yearbook: 1958*, 87. Missouri ranked last among the eight auger mining states in 1963; its single mine produced a minuscule 7,000 tons of coal. U.S. Bureau of Mines, *Minerals Yearbook: 1963*, 102–5.

35. Caudill, *Night Comes to the Cumberlands*, 310, 315.

36. C. L. Christensen, *Economic Redevelopment in Bituminous Coal: The Special Case of Technological Advance in the United States Coal Mines, 1930–1960* (Cambridge: Harvard University Press, 1962), 126–27, 252; Richard H. K. Vietor, *Environmental Politics and the Coal Coalition* (College Station: Texas A&M University Press, 1980), 15; U.S. Bureau of Mines, *Minerals Yearbook: 1948*, 279; U.S. Bureau of Mines, *Minerals Yearbook: 1958, Fuels*, 49, 141; U.S. Bureau of Mines, *Minerals Yearbook: Fuels, 1963*, 149, 175, 194; U.S. Bureau of Mines, *Minerals Yearbook: 1939*, 795; U.S. Bureau of Mines, *Mineral Yearbook: 1958*, 80–81; *United Mine Workers Journal* 65, no. 22 (November 1954): 12 (hereafter cited as *UMWJ*); *UMWJ* 67, no. 3 (February 1956): 10; *UMWJ* 71, no. 17 (September 1960): 17; U.S. Bureau of Mines, *Mineral Yearbook: 1963*, 91.

37. U.S. Bureau of Mines, *Minerals Yearbook: 1963*, 55; U.S. Bureau of Mines, *Minerals Yearbook: 1948*, 19; U.S. Bureau of Mines, *Minerals Yearbook: 1958*, 12; U.S. Bureau of Mines, *Minerals Yearbook: 1963*, 88–89.

Chapter Two

1. *Coal Age* 6 (December 1914): 965; Indiana, *Acts*, 1941, ch. 68, 172; Robert C. Meiners, "Strip Mining Legislation," *Natural Resources Journal* 3 (January 1964): 442–69.

2. Most histories of conservationism and environmentalism neglect to mention farmers at all, or they count them only as part of a backlash against governmental regulation. "To farmers, loggers, and miners, and to mining and manufacturing businesses," explains Richard Andrews, "nature was at best the ordinary raw material of economic commodities and livelihoods, and at worst a resistant or even threatening adversary, though as individuals they might also enjoy hunting and fishing." Richard N. L. Andrews, *Managing the Environment, Managing Ourselves: A History of American Environmental Policy* (New Haven: Yale University Press, 1999), 210. Richard Judd, however, has located the origins of conservation among nineteenth-century small farmers of northern New England. Richard Judd, *Common Lands, Common People: The Origins of Conservation in New England* (Cambridge: Harvard University Press, 1997).

3. *Report of the Strip Mining Study Commission to the Governor and the 97th General Assembly of the State of Ohio* (January 15, 1947), 10; *Coal Age* 9 (January 1916): 162.

4. Julian Feldman, "The Development of a Regulatory Policy for the Coal Stripping Industry in Ohio" (Master's thesis, Ohio State University, 1950), 28.

5. H. R. Moore and R. C. Headington, *Agricultural and Land Use As Affected by Strip Mining of Coal in Eastern Ohio* (Ohio State University Department of Rural Economics, Bulletin No. 135), 36, cited in Feldman, "The Development," 66; Charles Victor Riley, "An Ecological and Economic Study of Coal Stripped Land in Eastern Ohio" (Master's thesis, Ohio State University, 1947), 147; Mrs. James Angeloni to Governor C. William O'Neill, 6 March 1957, Box 33, Lausche Papers.

6. Riley, "An Ecological and Economic Study," 47–57.

7. *Cadiz Republican,* 13 September 1945, 1.

8. Feldman, "The Development," 85; *Report of the Strip Mining Study Commission,* 12; Jack K. Hill, "Social and Economic Implications of Strip Mining in Harrison County" (Master's thesis, Ohio State University, 1965), 61–62.

9. David Brooks, "Strip Mine Reclamation and Economic Analysis," *Natural Resources Journal* 6 (January 1966): 18–19; James Bristow, "Wildlife Havens from Strip Mines," *Outdoor America* (April–June 1943): 42–44; Feldman, "The Development," 86–87; *Report of the Strip Mining Study Commission,* 11–12.

10. *Coal Age* 40 (February 1935): 60–62; Feldman, "The Development," 99–100.

11. Feldman, "The Development," 100–106.

12. Ibid., 110–13; *Report of the Strip Mining Study Commission,* 6. Unfortunately, the testimony for the SMSC hearings has been lost and the arguments of the witnesses are available only second-hand, by way of the commission's *Report* and Julian Feldman's thesis.

13. *Report of the Strip Mining Study Commission,* 23–33.

14. *Cadiz Republican,* 30 January 1947, 1; Feldman, "The Development," 43, 45, 116–18; *Columbus Dispatch,* 6 June 1947, 1.

15. Ohio, *Laws,* 1947, p.730.

16. Feldman, "The Development," 118–20.

17. *Cadiz Republican,* 21 June 1945, 4.

18. Senator Lausche to Charles M. Perry, 14 February 1964, Box 165, Lausche Papers; *Cadiz Republican,* 30 August 1945, 1, and 25 October 1945, 1.

19. James Lent to Representative O'Neill, 19 May 1947, Folder "Strip Mining," Box 2, William O'Neill Papers, Ohio Historical Society (hereafter cited as O'Neill Papers).

20. *Journal of the Proceedings of the Ohio State Grange* (1943): 85 (hereafter cited as *JPOSG*); *JPOSG* (1946): 119; John R. Schofield to Governor Herbert, 10 June 1947, and H. D. Gerber to Governor Herbert, 25 May 1947, Folder "HB 314 to Regulate Strip Mining," Box 23, Thomas J. Herbert Papers, Ohio Historical Society (hereafter cited as Herbert Papers). County chapters of the Ohio Farm Bureau in support of H.B. 314 included: Ashtabula, Athens, Belmont, Columbiana, Delaware, Gallia, Guernsey, Harrison, Knox, Meigs, Morgan, Morrow, Muskingum, Perry, Pike, Portage, Richland, Ross, Seneca, Stark, Trumbull, Tuscarawas, and Washington.

21. Petition to Governor Herbert from John E. Thompson, 19 March 1947, and Fred S. Wheaton to Governor Herbert, 24 May 1947, Folder "HB 314 to Regulate Strip Mining," Box 23, Herbert Papers; Feldman, "The Development," 31; *Cadiz Republican,* 20 December 1945, 1; W. D. Matson editorial in letter from Milton Ronsheim to Governor Herbert, 13 May 1947, Folder "HB 314 to Regulate Strip Mining," Box 23, Herbert Papers.

22. *Cadiz Republican,* 7 February 1945, 1.

23. Ibid., 30 August 1945, 1; letter to the editor from Mrs. C. E. Householder, *Cadiz Republican,* 25 July 1945.

24. *JPOSG* (1945): 28–29.

25. Feldman, "The Development," 36; Western Union telegram from Ralph Abel & Son Lumber and Supply to Governor Herbert, 23 May 1947; Albert S. Adams to Governor Herbert, 22 May 1947; Western Union Telegram from R. A. Christian, The Can-

ton Supply Co., to Governor Herbert, 26 May 1947; Western Union Telegram from Gomer W. Jones, Jone's Restaurant, to Governor Herbert, 23 May 1947; Western Union Telegram from Fred Hoover to Governor Herbert, 23 May 1947; Western Union Telegram from Dale Brannon to Governor Herbert, 23 May 1947; Western Union Telegram from Bruns Coal Co., Inc., to Governor Herbert, 24 May 1947; and petition from R. S. Patterson, Beaver Fork Coal Company, to Governor Herbert, 23 May 1947, Folder "HB 314 to Regulate Strip Mining," Box 23, Herbert Papers.

26. Feldman, "The Development," 120–26; *Plain Dealer*, 10 April 1966.

27. Feldman, "The Development," 126–30.

28. Ibid., 135–36.

29. D. R. Stansfield to Senator Lausche, 17 April 1964, Box 165, Lausche Papers; *JPOSG* (1948): 131; *JPOSG* (1952): 103–4; *JPOSG* (1953); *JPOSG* (1960): 71–72; Ohio, *Laws*, 1959, 1231.

30. Ohio, *Acts*, 1965, Amendments to Section 1513, 495–96, 503; *Fed-O-Gram* (April 1965): 3; *Buckeye Farm News* (January 1969): 31.

31. *Times Leader*, 7 December 1970, 1; *Mountain Life and Work* 49 (January 1973): 18.

32. *Mountain Life and Work* 47 (February 1971): 15; Carole Malisiak to Ken Hechler, 24 August 1971, Folder 8, "Strip Mining," Box 169, Ken Hechler Manuscript Collection, Special Collections, Marshall University (hereafter cited as Hechler Papers).

Chapter Three

1. The few historians who have dealt with the subject of labor environmentalism have challenged the idea that an inherent clash of interests exists between workers and environmentalists. See Scott Dewey, "Working for the Environment: Organized Labor and the Origins of Environmentalism in the United States, 1948–1970," *Environmental History* 3 (January 1998): 45–46, as well as Robert Gordon, " 'Shell No!': OCAW and the Labor-Environment Alliance," *Environmental History* 3 (October 1998): 460–87, and "Poisons in the Fields: The United Farm Workers, Pesticides, and Environmental Politics," *Pacific Historical Review* 68 (February 1999): 51–77. This chapter and another article suggest that organized labor and environmentalists did not always have an untroubled, cooperative relationship; see Chad Montrie, "Expedient Environmentalism: Opposition to Coal Surface Mining in Appalachia and the United Mine Workers of America, 1945–1977," *Environmental History* 5 (January 2000): 75–98.

2. Pennsylvania, *Laws*, 1945, 1198, 1201, 1202.

3. *Dufour v. Maize*, 358 Pa. 309.

4. See Andrew Hurley on the ways in which Gary steelworkers' postwar recreational activities informed their environmental consciousness and activism. Andrew Hurley, *Environmental Inequalities: Class, Race, and Industrial Pollution in Gary, Indiana, 1945–1980* (Chapel Hill: University of North Carolina Press, 1995); Richard H. K. Vietor, *Environmental Politics and the Coal Coalition* (College Station: Texas A&M University Press, 1980), 59–60.

5. *Pittsburgh Press*, 21 June 1961, 1–2, and 25 June 1961, 1.

6. Ibid., 21 June 1961, 1–2; 25 June 1961, 1; 30 June 1961, 1; Vietor, *Environmental Politics*, 63.

7. *Pittsburgh Press*, 29 June 1961, 1; 4 July 1961, 1; 9 July 1961, 1.

8. Ibid., 10 July 1961, 1, 7.

9. Ibid., 11 July 1961, 6.

10. Ibid., 12 July 1961, 1.

11. Ibid., 16 July 1961, 1. Underscoring the implications of the Sunbeam president's remarks, an official with the state's Sanitary Water Board pointed out later in his testimony that Sunbeam had been cited on ten different occasions over a three-year period for discharging acid mine water into streams and thereby polluting a larger, major fishing stream.

12. Ibid., 26 July 1961, 1, 13.

13. Ibid., 2 August 1961, 1; 4 August 1961, 1, 2.

14. Ibid., 3 August 1961, 1; 5 August 1961, 1.

15. Ibid., 6 August 1961, 2, 3; 11 August 1961, 18; 7 August 1961, 2; 13 August 1961, 2.

16. Ibid., 8 August 1961, 1; 9 August 1961, 1; 16 August 1961, 1; 22 August 1961, 1.

17. Ibid., 23 August 1961, 1, 4, 16.

18. *Agreement between Bituminous Coal Stripping Association , Inc. and United Shovel Operators and Helpers Association, Inc., Bituminous District of Pennsylvania and State of Ohio, Effective April 1, 1937*, Box 65, United Mine Workers of America Papers, District #5, Indiana University of Pennsylvania (hereafter cited as UMW Papers); *Agreement between Operators of Strip Mines in Western Pennsylvania and United Mine Workers of America, District 5, Effective May 19, 1939*, Box 65, UMW Papers; *Agreement between the Operators of Strip Mines in Western Pennsylvania and Districts Nos. 3,4, and 5, UMWA, Effective April 29, 1941*, Box 65, UMW Papers; *United Mine Workers Journal* 61 (September 1950): 4 (hereafter cited as *UMWJ*).

19. *Pittsburgh Press*, 28 August 1961, 1, 7; 29 August 1961, 1, 2; Pennsylvania, *Laws*, 1961, 1210.

20. Vietor, *Environmental Politics*, 65–66; News Release, 5 October 1962, and News Release, 7 September 1962, Republican Campaign Headquarters, Box 8, File 23, "Conservation and Mining," William Scranton Papers, Historical Collections and Labor Archives, Pennsylvania State University (hereafter cited as Scranton Papers); *Pittsburgh Press*, 1 April 1962, 8; 9 September 1962, 1, 17.

21. *Pittsburgh Press*, 2 June 1963, 1; 5 June 1963, 1, 16.

22. Vietor, *Environmental Politics*, 67–69; *Pittsburgh Press*, 2 July 1963, 4; 9 July 1963, 1; 10 July 1963, 1, 10.

23. Ibid., 16 July 1963, 1; 2 June 1963, 16; 18 June 1963, 2; 6 June 1963, 2.

24. Vietor, *Environmental Politics*, 70–82.

25. Ibid., 17, 43–44; *UMWJ* 78 (March 1967): 4.

26. United Mine Workers of America, *Proceedings of the 45th Constitutional Convention* (United Mine Workers of America, 1968), 111, 162–63.

27. *UMWJ* 77 (May 1966): 8–9; *UMWJ* 77 (July 1966): 13; Senate Committee on Interior and Insular Affairs, *Hearings before the Committee on Interior and Insular Affairs on S. 3132, S. 3126, and S. 217, Bills to Provide for Federal-State Cooperation in the Reclamation of Strip Mined Lands*, 97, 113; *UMWJ* 79 (March 1968): 6.

28. Lewis Burke in Laurel Shackelford and Bill Weinberg, eds., *Our Appalachia* (New York: Hill and Wang, 1977), 353–54; *Mountain Life and Work* 48 (October 1972): 17.

29. The Yablonski campaign and other events leading up to an MFD victory are covered in greater detail in a number of sources. For the most thorough history, see Paul Nyden, "Miners for Democracy," (Ph.D. diss., Columbia University, 1974).

Chapter Four

1. Letter to the Editor from J. Vasas, *Pittsburgh Press*, 13 September 1962, 18.

2. Alice Slone, quoted in *To Save the Land and People*, 57 min., dir. Anne Lewis, Appalshop, Inc., 1999.

3. See Catherine McNicol Stock, *Rural Radicals: Righteous Rage in the American Grain* (Ithaca: Cornell University Press, 1996) and Pauline Maier, *From Resistance to Revolution: Colonial Radicals and the Development of Opposition to Britain, 1765–1775* (New York: Vintage, 1972).

4. Kentucky Strip Mining and Reclamation Commission, *Strip Mining in Kentucky* (Frankfort, 1965), 26; *Courier Journal*, 28 January 1966, 1, 14.

5. Kentucky, *Acts*, 1954, ch. 8., 20–21, 24; Marc Karnis Landy, *The Politics of Environmental Reform: Controlling Kentucky Strip Mining* (Washington, D.C.: Resources for the Future, 1976), 38; Charles Collier, "Influences of Strip Mining on the Hydrologic Environment of Parts of Beaver Creek Basin, Kentucky, 1955–59," U.S. Geological Survey Professional Paper 427-B; William Turner, "The Effects of Acid Mine Pollution on the Fish Population of Goose Creek, Clay County, Kentucky," *Progressive Fish-Culturist* 20 (January 1958): 45–46; *Courier Journal*, 17 January 1960, 18.

6. Lee Rose Pope Salyers to Harry Caudill, 7 September 1960, Folder 1, "Strip Mining, May 1956–Dec. 27, 1960," Box 19, Harry Caudill Papers, Special Collections, University of Kentucky (hereafter cited as Caudill Papers); E. V. Pope to Harry Caudill, 28 September 1960, Folder 1, "Strip Mining, May 1956–Dec. 27, 1960," Box 19, Caudill Papers; *Courier Journal*, 3 March 1962, 1, 4; *Mountain Eagle*, 1 March 1962; *Mountain Life and Work* 39 (1963): 59.

7. Harry Caudill to Mrs. Bingham, 19 April 1960, Folder 1, "Strip Mining, May 1956–Dec. 27, 1960," Box 19, Caudill Papers; *Mountain Eagle*, 28 July 1960, in Folder 1, "Strip Mining, May 1956–Dec. 27, 1960," Box 19, Caudill Papers.

8. Raymond Rash to Governor Combs, 13 September 1960, Folder 1, "Strip Mining, May 1956–Dec. 27, 1960," Box 19, Caudill Papers; Dexter Dixon to Governor Combs, 16 September 1960, Folder 1, "Strip Mining, May 1956–Dec. 27, 1960," Box 19, Caudill Papers.

9. Harry Caudill to Barry Bingham, 13 June 1962, Folder 2, "Strip Mining, Jan 7, 1961–Dec. 21, 1962," Box 19, Caudill Papers; Ronald Eller, *Miners, Millhands, and Mountaineers: Industrialization of the Appalachian South, 1880–1930* (Knoxville: University of Tennessee Press, 1982), 54. For an analysis of the motives behind this exchange of mineral rights see Robert S. Weise, *Grasping at Independence: Debt, Male Authority, and Mineral Rights in Appalachian Kentucky, 1850–1915* (Knoxville: University of Tennessee Press, 2001).

10. *Case v. Elkhorn Coal Corp.*, 210 Ky. 700, 276 S.W. 573, 574; *Rudd v. Hayden*, 265 Ky. 496, 97 S.W. 2d 35; *Treadway v. Wilson*, 301 Ky. 702, 192 S.W. 2d 949; Harry Caudill, *My Land Is Dying* (New York: E. P. Dutton, 1973), 63–65.

11. *Rochez Bros., Inc. v. Duricka*, Atl. 2d 825, 825–28; *Wilkes-Barre Township School District v. Corgan*, Atl. 2d 97, 97–99.

12. *Buchanan v. Watson*, 290 S.W. 2d 40, 40–43; David Schneider, "Strip Mining in Kentucky," *Kentucky Law Journal* 59 (1971): 654; Caudill, *My Land Is Dying*, 65. In the 1960 case, *Kodak Coal Company v. Smith*, the court of appeals upheld auger mining under a broad form deed, even when coal could be mined by other methods, arguing "if the operation is conducted properly the necessary use, or destruction of the surface is within the scope of the rights granted under the deed." *Kodak Coal Company v. Smith*, Ky. 338 S.W. 2d 699, 700.

13. In 1967, Pike County operators were offering payments of twenty-five cents per linear foot and royalties of ten cents per ton. Schneider, "Strip Mining in Kentucky," 655–56; *Courier Journal*, 5 January 1964, 11.

14. W. D. Bratcher to Harry Caudill, 25 September 1963, 14 November 1963, and 13 December 1963, Folder 3, "Strip Mining, March 17, 1963–November 19, 1964," Box 19, Caudill Papers. Although the state organization of soil conservation districts passed a resolution at its 1963 convention in support of new controls, the Letcher County district passed its own resolution calling for prohibition of contour and auger mining. Letcher County Soil Conservation District resolution, 13 December 1963, Folder 3, "Strip Mining, March 17, 1963–November 19, 1964," Box 19, Caudill Papers; Jewell Graham to Soil Conservation Supervisors of Kentucky, 3 February 1964, Folder 3, "Strip Mining, March 17, 1963–November 19, 1964," Box 19, Caudill Papers.

15. *Courier Journal*, 26 February 1964, 6; 17 March 1964, 8; 18 March 1964, 14; Kentucky, *Acts*, 1964, ch. 61, 233; Landy, *Politics of Environmental Reform*, 38–39.

16. Landy, *Politics of Environmental Reform*, 38–40, 124–35.

17. Bruce Daniel Rogers, "Public Policy and Pollution Abatement: TVA and Strip Mining" (Ph.D. diss., Indiana University, 1973), 8–9; David E. Whisnant, *Modernizing the Mountaineer: People, Power, and Planning in Appalachia* (Knoxville: University of Tennessee Press, 1994), 50; Caudill, *My Land Is Dying*, 69.

18. Rogers, "Public Policy and Pollution Abatement," 71, 79, 87, 103–6.

19. Ibid., 164–75. See the suggested legislation, "Appendix V," of Rogers's dissertation.

20. Mike Clark quoted in Guy and Candie Carawan, eds., *Voices from the Mountains* (New York: Alfred A. Knopf, 1975), 31; T. N. Bethell, "Hot Time Ahead," *Mountain Life and Work* (April) cited in *Appalachia in the Sixties: Decade of Reawakening*, ed. David S. Walls and John B. Stephenson (Lexington: University Press of Kentucky, 1972), 118–19; *Hazard Herald*, 10 April 1961 in Folder 2, "Strip Mining, Jan. 7, 1961–Dec. 21, 1962", Box 19, Caudill Papers; Dan Gibson and Bige Ritchie quoted in *People Speak Out on Strip Mining* (Berea, Ky.: Council of Southern Mountains, 1971), n.p.; Dan Gibson quoted in *To Look Over the Land and Take Care of It*, Broadside/Appalachian Video Network, Archives and Special Collections, East Tennessee State University.

21. *People Speak Out*; Caudill, *My Land Is Dying*, 76.

22. *Mountain Eagle*, 3 June 1965, 1–2.

23. Ibid., 10 June 1965, 1; 17 June 1965, 1; Harry Caudill to Richard Boone, 1 Novem-

ber 1965, Folder 4, "Strip Mining, January 10, 1965–December 23, 1965," Box 19, Caudill Papers; *New York Times*, 1 July 1965, 38.

24. Caudill, *My Land Is Dying*, 77–78; Landy, *Politics of Environmental Reform*, 134; *New York Times*, 23 June 1965, 63.

25. *Mountain Eagle*, 1 July 1965, 1; *Courier Journal*, 1 July 1965, 1, 20.

26. *Courier Journal*, 1 July 1965, 1, 20; Landy, *The Politics of Environmental Reform*, 135.

27. Defendants in the court action were Dan Gibson, Arch Engle, Tilves Ritchie, Alex Begley, Nathan Combs, Everett Combs, Herbert Combs, Elmer Williams, Delmer Williams, Clarence Williams, J. Garland Smith, Alfred Smith, Garland Smith, Bert Hollifield, Taylor Hurt, Matt Holliday, Cullin Ritchie, Bruce Ritchie, and Paul Ashley. *Mountain Eagle*, 22 July 1965, 12; *Courier Journal*, 22 July 1965, 1; 26 July 1965, 23.

28. Ibid., 27 July 1965, 1; 29 July 1965, 11, 26; 30 July 1965, 3.

29. *New York Times*, 4 July 1965, 27; 14 July 1965, 35; 1 August 1965, 36. See the "Surface Mining Reclamation and Conservation Requirements," in Rogers, "Public Policy and Pollution Abatement," Appendix III.

30. *Mountain Eagle*, 12 August 1965, 1; Landy, *Politics of Environmental Reform*, 135–37; Schneider, "Strip Mining in Kentucky," 660–61.

31. Dan Gibson quoted in *To Look Over the Land*; *Hazard Herald*, 15 November 1965, in Folder 4, "Strip Mining, January 10, 1965–December 23, 1965", Box 19, Caudill Papers; Caudill, *My Land Is Dying*, 80–81.

32. *New York Times*, 5 January 1966, 19; Landy, *Politics of Environmental Reform*, 40, 146–47.

33. Harry Caudill to Mr. Oz Johnson, 10 January 1966, Folder 5, "Strip Mining, January 5, 1966–March 24, 1966", Box 19, Caudill Papers.

34. Landy, *Politics of Environmental Reform*, 164–65; *Courier Journal*, 13 January 1966, 8; 9 January 1966, 27.

35. Landy, *Politics of Environmental Reform*, 171–72; *Courier Journal*, 19 January 1966, 1, 14; 21 January 1966, 1.

36. Kentucky, *Acts*, 1966, ch. 4; *Courier Journal*, 22 January 1966, 1; 26 January 1966, 1; 28 January 1966, 1; *New York Times*, 28 January 1966, 1. Even by 1971 there were fewer than twenty-five inspectors (who received little specialized training) carrying out 4,136 inspections on 22,000 acres of land, which included visiting every strip mine operation in the inspector's territory on a bimonthly basis. Landy, *Politics of Environmental Reform*, 189–90.

37. *Courier Journal*, 28 January 1966, 1; *New York Times*, 28 January 1966, 1; Landy, *Politics of Environmental Reform*, 41, 195–96.

38. Caudill, *My Land is Dying*, 81–82; *Martin v. Kentucky Oak Mining Co.*, 429 S.W. 2d 395; George Laycock, *The Diligent Destroyers* (New York: Doubleday, 1970), 163.

39. *Martin v. Kentucky Oak Mining Co.*, 429 S.W. 2d 395; *New York Times*, 15 January 1967, 64.

Chapter Five

1. Tony Dunbar, *Our Land Too* (New York: Pantheon, 1971), 138–39; *Strip Mining Bulletin*, no. 2, 26 June 1967 (Appalachian Group to Save the Land and People, Cordia,

Ky.), in Folder 1, "Strip Mining, Jan. 8, 1967–July 30, 1967," Box 20, Harry Caudill Papers, Special Collections, University of Kentucky (hereafter cited as Caudill Papers); Harlan County petition, Folder 9, "Stripmining, 1967–1968," Box 60, Records of the Appalachian Volunteers, 1963–1970, Southern Appalachian Archives, Berea College (hereafter cited as AV Papers).

2. *Daily Enterprise*, 7 July 1967, Folder 1, "Strip Mining, Jan. 8, 1967–July 30, 1967," Box 20, Caudill Papers.

3. "Strip Mining—Questions and Answers," Folder 6, Box 5, Gordon Ebersole/The Congress for Appalachian Development Manuscript Collection, East Tennessee State University (hereafter cited as Ebersole Papers).

4. "The Appalachian Group to Save the Land and People: What Kind of Action?," Folder 12, "Appalachian Group to Save the Land and People (AGSLP), 1967–1968," Box 40, AV Papers.

5. *New York Times*, 30 July 1967, 29; Harry Caudill, *My Land Is Dying* (New York: E. P. Dutton, 1973), 87; *Courier Journal*, 10 August 1967, 1. At the end of the summer of 1968, saboteurs destroyed $750,000 worth of equipment at a Round Mountain Coal Company operation in Leslie County (see Introduction). Later, in December, someone blew up $1 million worth of equipment, including one shovel and six bulldozers, at a Blue Diamond Coal strip operation in Campbell County, Tennessee, near the state border. T. N. Bethell, "Hot Time Ahead," *Mountain Life and Work* (April 1969), cited in *Appalachia in the Sixties: Decade of Reawakening*, ed. David S. Walls and John B. Stephenson (Lexington: University Press of Kentucky, 1972), 116–19.

6. Buck Maggard, quoted in *To Save the Land and People*, 57 min., dir. Anne Lewis, Appalshop, Inc., 1999; "What Kind of Action?," AV Papers.

7. The CSM executive board outlined its policy on strip mining in January, after receiving a letter from Harry Caudill asking the staff to write letters in support of the regulatory legislation before the General Assembly. The Council decided to have executive director Perle Ayer write a letter to the governor as well as to each of the state senators and representatives stating CSM's position "in favor of conservation practices." Take whatever action is necessary, he wrote, and at once, to ensure "fair and honest and guaranteed protection of each citizen against injustice to person, property, and future security and opportunity, which could otherwise occur as a direct result or related side effect to mining practices or other methods of harvesting national resources." The Council did not support or condemn specific legislation, Ayer explained, "but it does believe that this nation was founded upon and is still dedicated to justice and equal opportunity . . . defended as necessary against those who in short-sighted and self-interest do not recognize their departure from this principle." Memo from Milton Ogle to All Council Staff, 17 January 1966, and P. F. Ayer to Governor Breathitt, 18 January 1966, Folder 4, "Council of Southern Mountains: Correspondence and Minutes, 1966," Box 7, Ebersole Papers. On the reluctance of CSM to involve itself politically and the divisions that emerged at the end of the 1960s, see David Whisnant, "Controversy in God's Grand Division: The Council of the Southern Mountains," *Appalachian Journal* 2 (Autumn 1974): 7–45.

8. Interviews with Joe Mulloy (10 Nov. 1990) and Michael Kline (14 March 1991), The War on Poverty in Appalachian Kentucky Oral History Project, Special Collec-

tions, University of Kentucky (hereafter cited as War on Poverty Project); Harry Caudill to Stewart Udall, 9 August 1966, and 26 August 1966, Folder 6, "Strip Mining, April 1, 1966–December 28, 1966," Box 19, Caudill Papers. There were no SCEF activists in eastern Kentucky until mid-1966. Carl Braden to Harry Caudill, 25 January 1966, Folder 5, "Strip Mining, January 5, 1966–March 24, 1966," Box 19, Caudill Papers. The Hemp Hill residents were trying to save their water source and homes, most of which were purchased from the Elkhorn Coal Company in the 1950s, but they did not expect the petition to succeed. "We ought to just do like those old women did over in Knott County," an SCEF newsletter reported a Hemp Hill woman saying. "We ought to get our guns and go over there and tell 'em to git," she said. "I can shoot a .38." News From the Southern Conference Educational Fund, 14 September 1966, Folder 6, "Strip Mining, April 1, 1966–December 28, 1966," Box 19, Caudill Papers.

9. Interview with Joe Mulloy (10 Nov. 1990), War on Poverty Project; Tom Bethell to Milton Ogle, 4 September 1967, Folder 5, "Bethell, Tom, Director, AV Information Office, 1966–1969," Box 42, AV Papers; *New York Times*, 16 July 1967, 5; AGSLP press release, 13 July 1967, Folder 12, "Appalachian Group to Save the Land and People (AGSLP), 1967–1968," Box 40, AV Papers; Program from Strip Mining Symposium, 13 July 1967 (Owensboro: AV Information Office), Folder 6, Box 5, Ebersole Papers; "Why We Come to Owensboro: A Report to Governor Breathitt, 13 July 1967," Folder 6, Box 5, Ebersole Papers.

10. James C. Millstone, "East Kentucky Coal Makes Profits for Owners, Not Region," *St. Louis Post-Dispatch*, 18 and 20 November 1967, cited in *Appalachia in the Sixties*, 75; interview with Joe Mulloy (10 Nov. 1990), War on Poverty Project; *Courier Journal*, 4 July 1967, 1.

11. *Courier Journal*, 9 July 1967, 28; 11 July 1967, 20; *Mountain Eagle*, 20 July 1967, 1.

12. *Courier Journal*, 19 July 1967, 1; *New York Times*, 18 July 1967, 19; 19 July 1967, 18.

13. *Courier Journal*, 22 July 1967, 1; 2 August 1967, 20; 12 October 1967, 7; *Strip Mining Bulletin* (Special Island Creek Issue), 8 July 1967, Folder 1, "Strip Mining, Jan. 8, 1967–July 30, 1967," Box 20, Caudill Papers.

14. AGSLP Press Release, 24 July 1967, Folder 12, "Appalachian Group to Save the Land and People (AGSLP), 1967–1968," Box 40, Ebersole Papers; Morris Sheperd to Gordon Ebersole, 20 July 1967, Folder 6, Box 5, Ebersole Papers; *New York Times*, 30 July 1967, 29.

15. Gene L. Mason, "The Subversive Poor," *The Nation*, 30 December 1968, quoted in Caudill, *My Land Is Dying*, 88–89; interviews with Joe Mulloy and Karen Mulloy (10 Nov. 1990), War on Poverty Project.

16. *New York Times*, 27 August 1967, 71; *Courier Journal*, 27 September 1967, 1; Paul Good, "Kentucky's Coal Beds of Sedition," *The Nation*, 4 September 1967, in *Appalachia in the Sixties*, 192; interview with Joe Mulloy (10 Nov. 1990), War on Poverty Project.

17. Albert Whitehouse, Statement on Appalachian Volunteers Situation in Kentucky, 11 September 1967, p. 1, Folder 3, "Strip Mining, Sept. 1, 1967–Dec. 28, 1967," Box 20, Caudill Papers; Caudill, *My Land Is Dying*, 89; Calvin Trillin, "The Logical Thing, Costwise," *New Yorker*, 27 December 1969, cited in *Appalachia in the Sixties*, 113.

18. *Martin v. Kentucky Oak Mining Co.*, 429 S.W. 2d 395, 395–99. *Amicus curiae*

briefs were filed on behalf of the plaintiffs by David Schneider for the Commonwealth of Kentucky, Edward Post for the Kentucky Civil Liberties Union, James Young for the AGSLP, and Lawrence Grauman for the Sierra Club.

19. Ibid., 399–401; *Wiser Oil Company v. Conley, Ky.*, 346 S.W. 2d 718. The court cases from other states Hill cited included *Franklin v. Callicoat*, 119 N.E. 2d 688; *East Ohio Gas Company v. James Brothers Coal Company*, 85 N.E. 2d 816; *Williams v. Hay*, 120 Pa. 485, 14 Atl. 379; *Livingston v. Moingona Coal Company*, 49 Iowa 369, 21 Am. Rep. 150; *Carton v. So. Butte Mining Co.*, 181 Fed. 941, 104 C.C.A. 405,; *Oresta v. Romano Brothers*, 137 W.Va. 633, 73 S.E. 2d 622; *West Virginia-Pittsburgh Coal Company v. Strong*, 42 S.E. 2d 46; *Rochez Bros., Inc. v. Duricka*, 374 Pa. 262, 97 Atl. 2d 825; *C & O Railroad Company v. Bailey Production Corporation*, 163 F. Supp. 666; *Campbell v. Campbell, Tenn.*, 199 S.W. 2d 931); *United States v. Polino*, 131 F. Supp. 772, N.D. W.Va. 1955; *Wilkes-Barre Township School District v. Corgan*, 403 Pa. 383, 170 A. 2d. 97; *Rocky Mountain Fuel Co. v. Heflin*, 366 P. 2d; *Benton v. U.S. Manganese Corp.*, 313 S.W. 2d 839.

20. *Courier Journal*, 12 December 1967, 1.

21. Marc Karnis Landy, *The Politics of Environmental Reform: Controlling Kentucky Strip Mining* (Washington, D.C.: Resources for the Future, 1976), 250–67; Mathematica, Inc., *Design of Surface Mining Systems in the Eastern Kentucky Coal Fields*, vol. 1 (Frankfort: Kentucky Department of Natural Resources and Environmental Protection, 1974), 67 and A-10; Mathematica, Inc., *Design of Surface Mining Systems*, vol. 2, 10.

22. *Mountain Life and Work* 46 (October 1970): 3; *Courier Journal*, 10 October 1971, 1. Carter Combs was a former Michigan coal broker with neither a college degree nor the four years of conservation-related experience required of the inspectors he would be supervising. He received his appointment as supervisor of the northeast district— which included Pike County—because his daughter had been a campaign aide for Nunn. While working for the reclamation division, Combs did a great deal to diminish the quality of enforcement in the eastern Kentucky coalfield. Despite leading all other eastern Kentucky counties in coal production, Pike County had the lowest number of violations per inspection. Landy, *The Politics of Environmental Reform*, 282, 292–98.

23. Mathematica, *Design of Surface Systems*, vol. 1, 4, 33–36; Kenneth L. Dyer and Willie R. Curtis, *Effect of Strip Mining on Water Quality in Small Streams in Eastern Kentucky, 1967–1975*, USDA Forest Service Research Paper NE-372 (1977), 2, 10, 12.

24. Copy of form letter from H. A. Ritchie, 1969; KCC Resolution, October 17, 1969; and copy of form letter, 22 November 1969, Folder 1, "Kentucky Conservation Council and ad hoc Committee on Strip-Mining, Jan. 3, 1969–Dec. 31, 1969," Box 24, Caudill Papers.

25. Minutes of Meeting Called by KCC to Invite Other Organizations and Individuals to Determine Action on Strip Mining, Folder 1, "Kentucky Conservation Council and ad hoc Committee on Strip-Mining, Jan. 3, 1969–Dec. 31, 1969," Box 24, Caudill Papers.

26. Ibid.; Mike Flynn to Jim Butler, 18 December 1969, and Harry Caudill to Jim Butler, 18 December 1969, Folder 1, "Kentucky Conservation Council and ad hoc Committee on Strip-Mining, Jan. 3, 1969–Dec. 31, 1969," Box 24, Caudill Papers.

27. *Courier Journal*, 13 July 1967, 6; 7 June 1970, 1; *Mountain Life and Work* 46 (June

1970): 18; *New York Times*, 10 May 1970, 63, and 8 June 1970, 21; *SOK Newsletter* 2 (1971), Folder "Strip Mining," Boxes 1/2, West Virginia and Regional History Collection, West Virginia University (hereafter cited as WVRH Papers).

28. *Courier Journal*, 27 September 1970; Warren Wright, quoted in *People Speak Out on Strip Mining* (Berea, Ky: Council of Southern Mountains, 1971), n.p.

29. Jack Weller to Harry Caudill, 26 January 1971, Folder 4, "Kentucky Conservation Council and ad hoc Committee on Strip-Mining, Jan. 26, 1971–Sept. 27, 1971," Box 24, Caudill Papers. The Citizens' League to Protect Surface Rights was formed by citizens of Letcher County "fed up with the destruction of their land by strip mining and gas and oil drillings," primarily by the Kentucky-West Virginia Gas Company. Their complaints centered on the oil leaks, gas explosions, acid drainage, and landslides taking place in the creeks and hollows of Blackey. The league hoped "to halt further destruction and, where damage has already occurred, to guarantee just compensation for the landowner." The group's chair was Joe Begley, who had left employment by the gas company in West Virginia in 1966 to run his deceased father-in-law's store. See *Mountain Life and Work* 47 (January 1971): 22. The Appalachian Research and Defense Fund, also new to the scene in 1970, was formed when the AVs merged with the Mountain Legal Rights Organization and a group of lawyers in West Virginia.

30. *Mountain Life and Work* 47 (February 1971): 15; *New York Times*, 14 June 1971, 16; Minutes, Board of Directors, Save Our Kentucky, Inc., 8 August 1971, Folder 4, "Kentucky Conservation Council and ad hoc Committee on Strip-Mining, Jan. 26, 1971–Sept. 27, 1971," Box 24, Caudill Papers.

31. John Franson to Harry Caudill, 5 September 1970, Folder 3, "Kentucky Conservation Council and ad hoc Committee on Strip-Mining, June 9, 1970–Aug. 5, 1971," Box 24, Caudill Papers; *Mountain Life and Work* 45 (June 1969): 21; Harry Caudill to Mr. Harry LaViers Jr., 8 June 1971; John Franson to Harry Caudill, 12 July 1971; and John Franson to Dr. Elvis Stahr, 5 August 1971, Folder 3, "Kentucky Conservation Council and ad hoc Committee on Strip-Mining, June 9, 1970–Aug. 5, 1971," Box 24, Caudill Collection. A handwritten note on the last page of the third letter details this secret funding of SOK, and Caudill's papers at the University of Kentucky include a number of slips of deposit for $4,000 to Caudill's Bank of Whitesburg account.

32. *Mountain Life and Work* 47 (July–August 1971): 18; *Appalachian Strip Mining Information Service*, 13 November 1971, 3–5, Folder "Strip Mining," Boxes 1/2, WVRH Papers; *New York Times Magazine*, 12 December 1971, 116.

33. Legislative Research Commission, *Legislative Hearing: Surface Mining: Bulletin No. 94* (December 1971), 5–6, 8–9, 11, 28, 63, Folder 4, "Strip Mining Dec. 15, 1971–Dec. 31, 1971," Box 22, Caudill Papers.

34. Handbill and *SOK Bulletin*, Folder 5, "Save Our Kentucky, Inc., Oct. 2 1971–Dec. 1, 1971," Box 24, Caudill Papers; *Mountain Eagle*, 6 January 1972, 1, 14, 15.

35. *Courier Journal*, 21 January 1972; "We will stop the bulldozers—A Statement by the Women of Eastern Kentucky, January 20, 1972 Strip Mining," Folder, "Strip Mining," Boxes 1/2, WVRH Papers.

36. "We will stop the bulldozers"; *Courier Journal*, 21 January 1972.

Chapter Six

1. Barnard Aronson to the editor of the *Charleston Gazette*, n.d., n.p., in the private papers of Richard Cartwright Austin (hereafter cited as RCA Papers).

2. *Charleston Gazette*, 27 February 1971, 1; 11 February 1971, 20.

3. Robert F. Munn, "The First Fifty Years of Strip Mining in West Virginia, 1916–1965," *West Virginia History* 35 (October 1973): 66–74.

4. Thomas Sweeney is quoted by Robert Munn in the *Daily Herald* [Wellsburg, West Virginia], 2 February 1939, in "The First Fifty Years," 72.

5. West Virginia, *Acts*, 1939, ch. 84, 402.

6. Munn, "The First Fifty Years," 69; George Siems, "The Strip Mining of Bituminous Coal in West Virginia: An Analysis of Past and Present Conditions" (Master's thesis, University of Pennsylvania, 1949), 10; West Virginia, *Acts*, 1945, ch. 85, 347–48.

7. *Charleston Gazette*, 30 August 1958, 14.

8. Ellis Bailey quoted in *Citizens to Abolish Strip Mining, Inc.* [Charleston, West Virginia], February 1972, 2, 6, in "Strip Mining," Box 2, West Virginia and Regional History Collection, West Virginia University (hereafter cited as WVRH Papers).

9. *Charleston Gazette*, 2 February 1967, 6.

10. R. A. Schmidt and W. C. Stoneman, *A Study of Coal Surface Mining in West Virginia* (Menlo, Calif.: Stanford Research Institute, 1972), 108; *Charleston Gazette*, 24 January 1967, 15; 29 January 1967, 15; 30 January 1967, 2; 31 January 1967, 7; 1 February 1967, 1; 7 February 1967, 11; 8 February 1967, 1; 29 January 1967, 15.

11. *Charleston Gazette*, 1 February 1967, 1.

12. West Virginia, *Acts*, 1967–68, ch. 20, 914–15; *Charleston Gazette*, 18 February 1967, 1.

13. Schmidt and Stoneman, *A Study of Coal Surface Mining*, v, 40–42.

14. *Coal Age* (March 1971): 95.

15. *Charleston Gazette*, 19 January 1971, 2; Richard Cartwright Austin, interview with author, 15 September 1998, Dungannon, Virginia (tape recording and handwritten notes in possession of the author).

16. *Charleston Gazette*, 31 December 1970; News Release, 30 December 1970, Folder 4, "Strip Mining, Misc.," Box 177, Ken Hechler Manuscript Collection, Special Collections, Marshall University (hereafter cited as Hechler Papers).

17. *Charleston Gazette*, 1 January 1971; 5 January 1971; *Observer* [West Virginia AFL-CIO], March 1971, in Box 2, WVRH Papers.

18. *Charleston Gazette*, 6 January 1971, 3; 19 January 1971, 2; Richard Cartwright Austin, interview with author.

19. *Charleston Gazette*, 29 January 1971, 21; 2 February 1971, 4, 13; 11 February 1971, 30.

20. Ibid., 5 February 1971, 4; 10 February 1971, 20; 11 February 1971, 1.

21. Ibid., 16 January 1971, 4; 2 February 1971, 15; 9 February 1971, 3; 23 February 1971, 6.

22. Ibid., 3 March 1971, 19.

23. Ibid., 15 January 1971, 6; 21 January 1971, 1; 27 January 1971, 1; 29 January 1971, 20.

24. *Coal Age* (June 1971): 82.

25. *Charleston Gazette*, 16 February 1971, 1, 2.

26. Ibid., 18 February 1971, 1.

27. Ibid., 15 February 1971, 1; 20 February 1971, 22.

28. Ibid., 23 February 1971, 6; 2 March 1971, 1.

29. Ibid., 28 February 1971, 1.

30. Ibid., 15 February 1971, 1; 2 March 1971, 13; 4 March 1971, 37.

31. Ibid., 5 March 1971, 1; 6 March 1971, 1.

32. Ibid., 7 March 1971, 1; 10 March 1971, 1; 14 March 1971, 1; 18 March 1971, 1.

33. Schmidt and Stoneman, *A Study of Coal Surface Mining*, ii; "Coal Industry Connected with Stanford Stripping Investigation," *Citizens to Abolish Strip Mining, Inc.* (February 1972), 1, Folder "Strip Mining," Boxes 1/2, WVRH Papers.

34. Schmidt and Stoneman, *A Study of Coal Surface Mining*, 16, 42–44, 106–7, 118.

35. Jeanne Williams, "Strip Mining: In Some Ways Too Late for Boone," *Athenaeum* [West Virginia University], 31 March 1971, in RCA Papers. Member organizations in January 1972 included the Association for Environmental Protection, Inc. (Buckhannon); Boone County Citizens to Abolish Strip Mining (Madison); Citizens for Environmental Protection, Inc. (Charleston); Coal River Improvement Assoc., Inc. (St. Albans); Clay County Citizens to Abolish Strip Mining (Ivydale); Harrison County Concerned Citizens (Clarksburg); Marsh Fork Wildlife, Rod and Gun Club (Rock Creek); Montgomery Citizens Anti-Pollution Committee (Montgomery); Mountaineers Against Strip Mining (Charleston); Ohio Valley Citizens to Abolish Strip Mining (Parkersburg); West Virginia Highlands Conservancy (Charleston); West Virginia Public Affairs Conference (Charleston); West Virginia Union of Students (Morgantown); and Upper Monongahela Valley Citizens to Abolish Strip Mining (Morgantown). *Citizens to Abolish Strip Mining, Inc.* (January 1972) in Folder 6, "Miscellaneous," Box 176, Hechler Papers.

36. *Sierra Club Bulletin* 57 (May 1972): 21; Appalachian Strip Mining Information Service, 5/15/72, 2–3, and n.d./n.d./72, 2, Boxes 1/2, WVRH Papers.

37. *Mountain Life and Work* 47 (February 1971): 15; Carole Malisiak to Ken Hechler, 24 August 1971, Folder 8, "Strip Mining," Box 169, Hechler Papers; *Appalachian Strip Mining Information Service*, 10/n.d./71, 2, Folder "Strip Mining," Boxes 1/2, WVRH Papers.

Chapter Seven

1. Richard H. K. Vietor, *Environmental Politics and the Coal Coalition* (College Station: Texas A&M University Press, 1980), 16–18, 26.

2. U.S. Bureau of Mines, *Minerals Yearbook, 1973*, vol. 1, *Metals, Minerals, and Fuels*, 338; U.S. Bureau of Mines, *Minerals Yearbook, 1976*, vol. 1, *Metals, Minerals, and Fuels*, 363.

3. *Mining Congress Journal* 58 (April 1972): 77.

4. Perry E. Walper to John Saylor, 22 December 1958, Box 29, John P. Saylor Manuscript Collection, Special Collections, Indiana University of Pennsylvania (hereafter cited as Saylor Papers); John Saylor to Perry E. Walper, 12 January 1959, Box 29, Saylor Papers; John Saylor to Ross Leffler [assistant secretary of interior], 21 January 1959, Box 29, Saylor Papers; John Saylor to Perry E. Walper, 10 February 1959, Box 29, Saylor Papers; John Saylor to Representative Wayne Aspinall, 10 February 1959, Box 29,

Saylor Papers; John Saylor to Lewis E. Evans, 4 January 1961, Box 29, Saylor Papers; John Saylor to Stewart Udall, 2/23/61, Box 29, Saylor Papers; Charles Callison to John Saylor, 7/13/62, Box 29, Saylor Papers; Burton Martson to John Saylor, 8/12/63, Box 29, Saylor Papers.

5. *Courier Journal*, 30 September 1961, 1; 2 March 1962, 1, 16.

6. *Congressional Record*, 87th Cong., 2d sess., 1962, 8482; Robert C. Meiners, "Strip Mining Legislation," *Natural Resources Journal* 3 (January 1964): 460–63.

7. Donald P. Tonty to Frank Lausche, 10 July 1963, Box 156, Frank J. Lausche Papers, Ohio Historical Society (hereafter cited as Lausche Papers); Brian Winters and John Tozer to Frank Lausche, 28 June 1963, and John Tozer to Frank Lausche, 10 July 1963, Box 156, Lausche Papers; William A. Riaski to Frank Lausche, 17 November 1964, Box 156, Lausche Papers.

8. President's Appalachian Regional Commission, *Appalachia*, 44; David E. Whisnant, *Modernizing the Mountaineer: People, Power, and Planning in Appalachia* (Knoxville: University of Tennessee Press, 1994 [1980]), 127; George Laycock, *The Diligent Destroyers* (New York: Doubleday, 1970), 147–48.

9. U.S. Department of Interior, *Study of Strip and Surface Mining in Appalachia: An Interim Report to the Appalachian Regional Commission*, 22–24. Measuring a slope by degrees is based upon a 360-degree circle. A shear vertical wall would be one-fourth of a circle, or 90 degrees. A slope that rises as quickly as it proceeds along the horizontal is a 45-degree slope. See Mark Squillace, *The Strip Mining Handbook: A Coalfield Citizens Guide to Using the Law to Fight Back against the Ravages of Strip Mining and Underground Mining* (Washington, D.C.: Environmental Policy Institute and Friends of the Earth, 1990), 101.

10. *Congressional Record*, 89th Cong., 2d sess., 1966, 18557; U.S. Department of Interior, *Study of Strip and Surface Mining in Appalachia*, 4–7.

11. *New York Times*, 16 July 1967, 5; 25 July 1967, 34; Donald McIntosh to Frank Lausche, 16 December 1966, Box 187, Lausche Papers; Robert Ritchie to Frank Lausche, 14 December 1966, Box 187, Lausche Papers; *Congressional Record*, 89th Cong., 2d sess., 1966, 25,176–77.

12. Senate Committee on Interior and Insular Affairs, *Hearings before the Committee on Interior and Insular Affairs on S. 3132, S. 3126, and S. 217, Bills to Provide for Federal-State Cooperation in the Reclamation of Strip Mined Lands*, 1–2, 4–7, 11–13, 19–21.

13. Senate Committee on Interior and Insular Affairs, *Hearings before the Committee*, 34–40, 97, 113,

14. Harry Caudill to Bob Waldrup, 1 May 1968, Folder 4, "Strip Mining, Jan. 3, 1968–Nov. 30, 1968," Box 20, Harry Caudill Manuscript Collection, Special Collections, University of Kentucky (hereafter cited as Caudill Papers); Collection Guide, Gordon Ebersole/The Congress for Appalachian Development Manuscript Collection, East Tennessee State University (hereafter cited as Ebersole Papers); Johanna Henn to Harry Caudill, 3 April 1967, Folder 1, "Strip Mining, Jan. 8, 1967–July 30, 1967," Box 20, Caudill Papers; Harry Caudill to Johanna Henn, 8 April 1967, Folder 1, "Strip Mining, Jan. 8, 1967–July 30, 1967," Box 20, Caudill Papers; James Kowalsky to Harry Caudill, 24 October 1967, Folder 3, "Strip Mining, Sept. 1, 1967–Dec. 28, 1967,"

Box 20, Caudill Papers; Harry Caudill to Tom Bethell, 21 November 1967, Folder 3, "Strip Mining, Sept. 1, 1967–Dec. 28, 1967," Box 20, Caudill Papers.

15. Senate Committee on Interior and Insular Affairs, *Hearings before the Committee*, 87–91.

16. Ibid., 283–85; statement by Alice Grossniklaus to Frank Lausche, 26 April 1963, Box 156, Lausche Papers.

17. Senate Committee on Interior and Insular Affairs, *Hearings before the Committee*, 1968, 298, 302, 337–38, 342–43, 365.

18. *New York Times*, 22 December 1970, 32.

19. House Subcommittee on Mines and Mining, *Hearings before the Subcommittee on Mines and Mining of the Committee on Interior and Insular Affairs . . . on H.R. 60 and Related Bills, Relating to Strip Mining*, 16–23.

20. Petitions to Ken Hechler, 7 January 1971, Folder 6, "Strip Mining," Box 170, Ken Hechler Manuscript Collection, Special Collections, Marshall University (hereafter cited as Hechler Papers).

21. Ken Hechler, press releases, 11 February 1971, and 18 February 1971, copies in author's possession.

22. Ibid., 22 April 1971, 21 May 1971, copies in author's possession; announcement, 22 April 1971, Folder 6, "Strip Mining," Box 170, Hechler Papers.

23. *United Mine Workers Journal* 82 (January 1971): 4, 6 (hereafter cited as *UMWJ*); *UMWJ* 82 (February 1971): 3; *UMWJ* 82 (March 1971): 9.

24. *Sierra Club Bulletin* 56 (February 1971): 7; *Mining Congress Journal* 57 (September 1971): 138.

25. *Appalachian Strip Mining Information Service Bulletin* (10/n.d./71), 2, Folder "Strip Mining," Boxes 1/2, WVRH Papers. NCASM was formed sometime between the end of October 1971 and January 1972.

26. *Appalachian Strip Mining and Information Service*, 10/8/71, 2, Folder "Strip Mining," Boxes 1/2, West Virginia and Regional History Collection, West Virginia University (hereafter cited as WVRH Papers). The cover page of each bulletin listed the organizations being serviced. In March 1972 this included End Environmental and Social Devastation from Coal Strip Mining, Environmental Policy Center, Friends of the Earth, national Coalition Against Strip Mining, Sierra Club; Regional: Appalachian Coalition, Council of the Southern Mountains, Izaak Walton League (South East Region); Arizona: Black Mesa Defense Fund; Kentucky: Save Our Kentucky; Montana: Bull Mountain Landowners Association; Ohio: Committee to Control Strip Mining, Stop Ohio Stripping; Pennsylvania: Help Eliminate Life Pollutants; Tennessee: Save Our Cumberland Mountains, Tennessee Citizens for Wilderness Planning; Virginia: Wise County Environmental Council; West Virginia: Citizens to Abolish Strip Mining, Highlands Conservancy, Izaak Walton League. *Appalachian Strip Mining and Information Service*, 3/14/72, 1, Folder "Strip Mining," Boxes 1/2, WVRH Papers; *Appalachian Strip Mining and Information Service*, 11/13/71, 2, Folder "Strip Mining," Boxes 1/2, WVRH Papers; Austin put out the last issue in either late October or early November, 1972. *Appalachian Strip Mining and Information Service*, n.d./n.d./72, 1, Folder "Strip Mining," Boxes 1/2, WVRH Papers.

27. House Subcommittee on Mines and Mining, *Hearings before the Subcommittee on Mines and Mining . . . on H.R. 60*, 165–72, 638; Senate Subcommittee on Minerals, Materials, and Fuels, *Hearings before the Subcommittee on Minerals, Materials, and Fuels of the Committee of Interior and Insular Affairs Pursuant to S. Res. 45, A National Fuels and Energy Policy Study, on S. 77 . . . S. 2777*, 151–56.

28. Ken Hechler, "Notes on Testimony on Strip Mining, September 20, 1971," copies in author's possession.

29. Ibid.

30. Ibid.

31. Ibid.

32. Senate Subcommittee on Minerals, Materials, and Fuels, *Hearings before the Subcommittee . . . on S. 77*, 280–82, 329–32, 620; House Subcommittee on Mines and Mining, *Hearings before the Subcommittee . . . on H.R. 60*, 296, 559, 562–63.

33. *UMWJ* 82 (February 1971): 3; *UMWJ* 82 (March 1971): 9; *UMWJ* 82 (October 1971): 24; Ken Hechler, press release, 21 May 1971, copy in author's possession. President John Lewis first demanded that coal operators establish a welfare and retirement fund, financed by a per-ton royalty, in 1946. The operators refused, the miners went out on strike, and the government seized the mines. For a year the mines were operated under a contract that included a provision for a welfare fund, funded by a five-cent-per-ton royalty. The government returned the mines to the owners in June 1947, and the coal operators agreed to a contract that preserved the welfare fund but increased the royalty to ten cents per ton; *UMWJ* 82 (November 1971): 7–8, 10; Senate Subcommittee on Minerals, Materials, and Fuels, *Hearings before the Subcommittee . . . on S. 77*, 455, 458, 473; *UMWJ* 82 (May 1972): 24.

34. Senate Subcommittee on Minerals, Materials, and Fuels, *Hearings before the Subcommittee . . . on S. 77*, 224–27, 231; House Subcommittee on Mines and Mining, *Hearings before the Subcommittee . . . on H.R. 60*, 540–43.

35. House Subcommittee on Mines and Mining, *Hearings before the Subcommittee . . . on H.R. 60*, 549–50.

36. Ibid., 690–91, 711–12.

37. Ibid., 500; Senate Subcommittee on Minerals, Materials, and Fuels, *Hearings before the Subcommittee . . . on S. 77*, 1032–35.

38. House Subcommittee on Mines and Mining, *Hearings before the Subcommittee . . . on H.R. 60*, 496; Senate Subcommittee on Minerals, Materials, and Fuels, *Hearings before the Subcommittee . . . on S. 77*, 1032–35; Louise Dunlap to John Saylor, 11 October 1972, Box 29, Saylor Papers. FOE split in January and many of its former lobbyists left to form the EPC that same month. Louise Dunlap and Joe Browder waited until the end of January to resign from FOE and began representing the EPC in mid-February. The new group was formed, according to Dunlap, "to provide a much needed organization structure whose primary purpose and commitment will be environmental lobbying and litigation." Environmental Policy Center to Friends and Conservation Leaders, 27 January 1972, and Louise Dunlap to Harry Caudill, 1 February 1972, Folder 1, "Strip Mining Jan. 4, 1972–Apr. 27, 1972," Box 23, Caudill Papers.

39. House Subcommittee on Mines and Mining, *Hearings before the Subcommittee . . . on H.R. 60*, 373–77; Senate Subcommittee on Minerals, Materials, and Fuels, *Hear-*

ings before the Subcommittee . . . on S. 77, 489; *Sierra Club Bulletin* 56 (May 1971): 8; *Sierra Club Bulletin* 57 (May 1972): 20.

40. *People Speak Out on Strip Mining* (Berea, Ky.: Council of Southern Mountains, 1971), n.p.

41. *Mountain Life and Work* 48 (May 1972): 15; *Mountain Life and Work* 48 (June–July 1972): 23. Sources do not indicate what happened at the strip mine after this action.

42. Fred Harris to Keith Dix, 9 June 1972, Folder "Strip Mining," Boxes 1/2, WVRH Papers; *Mountain Life and Work* 48 (June–July 1972): 20, 26.

43. Memo from Louise Dunlap to Environmental Leaders and Resolution in *Appalachian Strip Mining Information Service Bulletin* (13 July 1972), Folder "Strip Mining," Boxes 1/2, WVRH Papers.

44. The Buffalo Creek disaster occurred early on the morning of February 26, when a Pittston dam constructed of mine refuse washed away. The dam was used to impound water for use at a tipple, where both deep-mined and stripped coal was washed. Its collapse sent 20 million cubic feet of water, mud, rock, and coal wastes through the valley below. Concern about the dam had prompted some nearby families to take refuge in a schoolhouse 5 miles down the creek the night before. But nearly five thousand people lived in the path of the flood and the waters destroyed sixteen communities and took more than 120 lives. Memo from Louise Dunlap to Environmental Leaders and Resolution in *Appalachian Strip Mining Information Service Bulletin* (13 July 1972), Folder "Strip Mining," Boxes 1/2, WVRH Papers; Senate Committee on Interior and Insular Affairs, *Surface Mining Reclamation Act of 1972: Report . . . to Accompany S. 630*; House Committee on Interior and Insular Affairs, *Report to Accompany H.R. 6482*.

45. Memo from Louise Dunlap and Environmental Leaders.

46. *Charleston Gazette*, 5 February 1971, 4; George W. Hopkins, "The Wheeling Convention of the Miners for Democracy: A Case Study of Union Reform Politics in Appalachia," in *Appalachia/America* (Johnson City: East Tennessee State University, 1980), 11; United Mine Workers of America, *Proceedings of the 46th Constitutional Convention, 1973* (United Mine Workers of America, 19??), 298.

47. *UMWJ* 83 (July 1972): 23, 27; *UMWJ* 83 (September 1972): 21; *Mountain Life and Work* 48 (June–July 1972): 16; *UMWJ* 83 (October 1972): 19.

48. Paul Nyden, "Miners for Democracy: Struggle in the Coal Fields" (Ph.D. diss., Columbia University, 1974), 479, 746, 838; *UMWJ* 83 (December 1972): 3.

Chapter Eight

1. Senate Committee on Interior and Insular Affairs, *Hearings before the Committee on Interior and Insular Affairs on S. 425 . . . [and] S. 923*, 1338–42, 1367–68; House Subcommittee on the Environment and Subcommittee on Mines and Mining, *Hearings before the Subcommittee on the Environment and Subcommittee on Mines and Mining of the Committee on Interior and Insular Affairs on H.R. 3 and Related Bills, Relating to the Regulation of Surface Mining*, 976, 1316–19.

2. House Subcommittee, *Hearings before the Subcommittee . . . on H.R. 3*, 1434–39;

Senate Committee on Interior and Insular Affairs, *Hearings before the Committee . . . on S. 425*, 441.

3. Senate Committee, *Hearings before the Committee . . . on S. 425*, 397–98; *UMWJ* 84 (April 1973): 17; *UMWJ* 84 (April 1973): 14, 19.

4. "Message from Congressman Ken Hechler," 26 January 1973, copy in author's possession; House Subcommittee, *Hearings before the Subcommittee . . . on H.R. 3*, 171–84, 775, 783–84; Senate Committee, *Hearings before the Committee . . . on S. 425*, 107; *Congressional Record*, 93rd Cong., 2d sess., 1973, 42966–72.

5. Memo from B. Lloyd and Jim Coen to "Appalachian Coalition People," Folder "Strip Mining—'73," private papers of Joe and Gaynell Begley (hereafter cited as Begley Papers); Baldwin Lloyd to Ken Hechler, 25 July 1973, Folder 5, "Strip Mining," Box 169, Ken Hechler Manuscript Collection, Special Collections, Marshall University (hereafter cited as Hechler Papers). In 1973, Lloyd was president of the Appalachian Coalition, J. W. Bradley was vice president, and there were a number of "state coordinators": James L. Coen Jr. in Virginia, Dr. Wayne C. Spiggle in Maryland, Richard J. Garrett in Ohio, J. W. Bradley in Tennessee, Robert Handley in West Virginia, and Joe Begley in Kentucky. Member organizations included CASM, SOCM, SOK, Citizens Committee to Investigate Strip Mining (Virginia), and Citizens Organized to Defend the Environment (Ohio).

6. House Subcommittee, *Hearings before the Subcommittee . . . on H.R. 3*, 1326, 1333; M. J. Clark and Rev. R. B. Lloyd, "Is Strip Mining Obscene?: A Case for the Abolition of Strip Mining in Appalachia," Vantage Point, vol. 5, no. 7, Folder "Strip Mining—'72," Begley Papers.

7. House Subcommittee, *Hearings before the Subcommittee . . . on H.R. 3*, 1110, 1119, 1642; Stephen Bossi to John Saylor, 25 May 1973, Folder "Correspondence Letters," Box 28, John P. Saylor Manuscript Collection, Special Collections, Indiana University of Pennsylvania (hereafter cited as Saylor Papers). Although there are a number of translations, Psalm 24:1 can be read as "The earth is the Lord's and all it holds, the world and those who live there." *The New American Bible* (Canada: World Catholic Press, 1987), 560.

8. House Subcommittee, *Hearings before the Subcommittee . . . on H.R. 3*, 1265, 1270, 1345, 1618–19; *Mountain Life and Work* 49 (March 1973): 2, 23.

9. House Subcommittee, *Hearings before the Subcommittee . . . on H.R. 3*, 1610.

10. Ibid., 932–34; *Reporter* 3, Special Issue (August 1997): 12–13; Senate Committee, *Hearings before the Committee . . . on S. 425*, 894–96.

11. Senate Committee, *Hearings before the Committee . . . on S. 425*, 911–15; House Subcommittee, *Hearings before the Subcommittee . . . on H.R. 3*, 1645–46; Sierra Club, *The Strip Mining of America* (1973), quoted in F. W. Schaller and P. Sutton, eds., *Reclamation of Drastically Disturbed Lands* (Madison, Wis.: American Society of Agronomy, 1978), 82.

12. Sierra Club, *Strip Mining of America*, 901; House Subcommittee, *Hearings before the Subcommittee . . . on H.R. 3*, 958, 969.

13. Ned Helmes to Ken Hechler, 30 March [1973], Folder 7, "Strip Mining," Box 170, Hechler Papers.

14. Senate Committee on Interior and Insular Affairs, *Surface Mining Reclamation Act of 1973, Report to Accompany S. 425*, 4–18.

15. Coalition Against Strip Mining, "Notice: Legislative Strip Mine Meeting, Washington, 1/26, 27, 28/1974," 2 January 1974, Folder 7, "Strip Mining," Hechler Papers; "Persons Attending EPC Sponsored Conference on Strip Mining Legislation," Folder 6, "Strip Mining," Box 170, Hechler Papers. Although there is no specific date for the memo sent by Baldwin and Johnson, the memo itself notes that the meeting had already passed. B. Lloyd and Linda Johnson to Appalachian Coalition People, n.d., Folder 6, Box 5, Gordon Ebersole/The Congress for Appalachian Development Manuscript Collection, East Tennessee State University (hereafter cited as Ebersole Papers). The Helmes memo must have been sent before May because its contents suggest that H.R. 11500 was still in committee. Ned Helmes to Ken Hechler, n.d., Folder 7, "Strip Mining," Box 170, Hechler Papers. In late January, Arizona congressman Morris Udall, supposedly an ally of the strip mining opposition, told activists that even getting a weak H.R. 11500 passed would be difficult. "[W]e'll be lucky to get this much through committee," he said, "and we might even lose a lot in committee or on the floor. So please, don't talk to me about making the bill any stronger." *Mountain Life and Work* 50 (February 1974): 8.

16. Jack Beidler to Ken Hechler, 12 July 1974, Folder 11, "Strip Mining," Box 170, Hechler Papers; A. F. Gospiron to Ken Hechler, 16 July 1974, Folder 11, "Strip Mining," Box 170, Hechler Papers; *Mountain Life and Work* 50 (July–August 1974): 18; *UMWJ* 85 (July 1974): 5. Unfortunately, the United Mine Workers of America will not allow researchers access to the minutes for this important board meeting.

17. *Mountain Life and Work* 50 (February 1974): 9; J. W. Bradley to Ken Hechler, 8 July 1974, Folder 9, "Strip Mining," Box 170, Hechler Papers; Wilburn C. Campbell to Ken Hechler, 5 July 1974, Folder 13, "Strip Mining," Box 170, Hechler Papers.

18. Charles E. Crank Jr. to Ken Hechler, 12 July 1974, Folder 11, "Strip Mining," Box 168, Hechler Papers; Jean Jones to Ken Hechler, 15 July 1974, Folder 2, "Strip Mining," Box 169, Hechler Papers; Patrick Neal to Ken Hechler, reply dated 2 August 1974, Folder 10, "Strip Mining," Box 169, Hechler Papers; Alice H. Slone to Ken Hechler, 25 June 1974, Folder 1, "Strip Mining," Box 170, Hechler Papers.

19. Ken Hechler, press release, 8 July 1974, copy in author's possession; Ken Hechler, press release, 12 July 1974, copy in author's possession.

20. Congressmen Udall opposed the surface owner consent provision. While protecting the rights of surface owners was important, he said, the committee bill's approach would give the surface owner "large windfall benefits from property that he does not own" through the exercise of veto power over "the right to mine someone else's coal". Udall was primarily concerned largely with the situation in the West, where many surface owners had never owned the mineral rights that still belonged to the federal government. He also acknowledged the need to compensate surface owners for any damage to their land. House Committee on Interior and Insular Affairs, *Surface Mining Control and Reclamation Act of 1974, Report . . . to Accompany H.R. 11500*, 4–19, 34–35, 46, 60–61, 179–80.

21. Coalition Against Strip Mining, 8 July 1974, Folder 7, "Strip Mining," Box 170,

Hechler Papers; John Melcher to Ken Hechler, 6 December 1974, Folder 9, "Strip Mining," Box 170, Hechler Papers; *New York Times,* 14 December 1974, 1, 14, and 18 December 1974, 22.

22. Ken Hechler to "Friends of the Coalition Against Strip Mining," 18 January 1975, Folder 8, "Strip Mining," Box 170, Hechler Papers.

23. *SOCM Sentinel* (January 1975): 1–2; Stephen Carl Cawood to Ken Hechler, 13 May 1975, Folder 11, "Strip Mining," Box 168, Hechler Papers; Coalition Against Strip Mining, January 1975, Folder 7, "Strip Mining," Box 170, Hechler Papers.

24. *World-News,* 9 April 1975, 14 April 1975; *Movin' On to Washington,* Broadside/ Appalachian Video Network, Archives and Special Collections, East Tennessee State University.

25. House and Senate Committee of Conference, *Surface Mining Control and Reclamation Act of 1975: Report . . . to Accompany H.R. 25,* 76, 87; Senate Committee on Interior and Insular Affairs, *Committee on Interior and Insular Affairs, Surface Mining Control and Reclamation Act of 1975: Report . . . to Accompany S. 7,* 149–50.

26. House Subcommittee on Energy and the Environment and Subcommittee on Mines and Mining, *Hearings Before the Subcommittee on Energy and the Environment and the Subcommittee on Mines and Mining of the Committee on Interior and Insular Affairs on the President's Veto of H.R. 25,* 13–17.

27. Ken Hechler, press release, 20 May 1975, copy in author's possession; Patricia Longfellow to Ken Hechler, 5 June 1975, Folder 5, "Strip Mining," Box 169, Hechler Papers. Citizens for Social and Economic Justice originated in the strip coalfields of southwestern Virginia as a Welfare Rights Organization but then broadened its scope to include other issues. By the fall a chapter had been organized in Knott County and essentially replaced AGSLP as the primary vehicle for grassroots struggle against strip mining. Within a couple of years, CESJ had chapters with thousands of members in other eastern Kentucky counties, eastern Tennessee, and West Virginia. Mart Shepard to Ken Hechler, [5 June 1975], Folder 15, "Strip Mining," Box 169, Hechler Papers, and *Mountain Life and Work* 52 (February 1976): 3; Charles M. Douglas to Ken Hechler, 29 May 1975, Folder 9, "Strip Mining," Box 169, Hechler Papers; Ken Hechler, press release, 13 June 1975, copy in author's possession.

28. "Declaration of Policy" in *Mining Congress Journal* 61 (November 1975): 89.

29. *Mountain Life and Work* 52 (October 1976): 41; United Mine Workers of America, *Proceedings of the 47th Constitutional Convention* (United Mine Workers of America, 1976), 273–76.

30. *SOCM Sentinel* (November 1975): 9; *SOCM Sentinel* (February 1976): 10, 11; *SOCM Sentinel* (March 1976): 2; *SOCM Sentinel* (May 1976): 6; *Mountain Life and Work* 52 (March and April 1976): 34.

31. House Committee on Interior and Insular Affairs, *Surface Mining Control and Reclamation Act of 1976: Report . . . to Accompany H.R. 9725,* 129; House Committee on Interior and Insular Affairs, *Surface Mining Control and Reclamation Act of 1976: Report . . . to Accompany H.R. 13950.*

32. House Subcommittee on Energy and the Environment, *Hearings before the Subcommittee on Energy and the Environment of the Committee on Interior and Insular*

Affairs on H.R. 2, Surface Mining Control and Reclamation Act of 1977, 37–38, 185 (Part I).

33. Ibid., 77, 10–12, 91–94 (Part I).

34. Ibid., 43 (Part II); *Mountain Life and Work* 53 (February–March 1977): 9; *Mountain Life and Work* 53 (April 1977): 42; *Reporter*, Special Issue, 3 August 1997, 7.

35. House Subcommittee on Energy and the Environment, *Hearings before the Subcommittee . . . on H.R. 2*, 30–31, 197 (Part III). Strip mining opponents established Save Our Mountains (SOM) in the early part of 1975. Initially it was a coalition of nine local groups designed to link together their resistance efforts. "We hope to stay away from what the anti-stripmining movement has been before," said first vice president Wayne Coombs, "and have grass roots groups and individuals involved." The stated goals of the coalition were abolition of strip mining, "gob" pile reclamation, freedom of information legislation, and stronger open meeting laws. SOM also intended to emphasize the need for jobs for miners in the event of a phaseout, but for most of 1975 the group was caught up in a campaign to prevent strip mining on 690 acres of Shavers' Fork. In January 1976, SOM began working to pass an abolition bill through the state legislature and lobbying against a bill that would remove counties from the moratorium on surface coal mining. But it was still doing little if anything to address the issues of job loss and economic impact caused by a phaseout or stricter regulations. *Mountain Life and Work* 50 (April 1975): 30; *Mountain Life and Work* 50 (December 1975): 10; *Mountain Life and Work* 52 (January 1976): 14.

36. House Subcommittee on Energy and the Environment, *Hearings before the Subcommittee . . . on H.R. 2*, 9 (Part I); *Mountain Life and Work* 52 (November 1976): 30–31. Virginia enacted its first control law in 1966, which it weakened considerably with coal industry-backed amendments in 1972, and subsequent efforts to move a stronger bill through the state legislature had failed. VCBR was established in April 1976 to rectify the inadequacies of strip mine control legislation and inspection, and it had a nine-member board of directors as well as membership of 250 by the time Kilgore appeared before the House committee in 1977.

37. House Subcommittee on Energy and the Environment, *Hearings before the Subcommittee . . . on H.R. 2*, 25–26 (Part III), 521–22 (Part IV).

38. Ibid., 93 (Part IV).

39. *New York Times*, 20 April 1977, 19; *Courier Journal*, 1 July 1977, 1, 8.

40. *Courier Journal*, 1 July 1977, 8; Senate Committee on Interior and Insular Affairs, *Committee on Interior and Insular Affairs, Surface Mining Control and Reclamation Act of 1977: Conference Report . . . to Accompany H.R. 2*, 98, 101, 104–06, 115.

41. *Mountain Life and Work* 53 (August 1977): 33–34.

42. *New York Times*, 4 August 1977, 1, 17; *Reporter*, Special Issue, 3 August 1997, 7; *Mountain Life and Work* 53 (August 1977): 31–32.

43. *Mountain Life and Work* 53 (October 1977): 26. William Eichbaum, formerly Solicitor in Pennsylvania, filled the post of deputy solicitor (head of enforcement) at OSM. *SOCM Sentinel* (April 1978): 8; *Washington Post*, 24 September 1978, in East Tennessee State University Vertical File; *Mountain Life and Work* 54 (April 1978): 29; *SOCM Sentinel* (August 1978): 9; Senate Subcommittee on Public Lands and Resources, *Hearing before the Subcommittee on Public Lands and Resources of the Com-*

mittee on Energy and Natural Resources on the Implementation of Public Law 95-87, the Surface Mining Control and Reclamation Act and Pending Legislation to Increase Authorization of Appropriations.

44. Senate Subcommittee on Public Lands, *Hearing*, 11–12; *SOCM Sentinel* (April 1978); *Mountain Life and Work* 54 (July 1978): 6.

45. *Reporter*, Special Issue, 3 August 1997, 14–15.

Chapter Nine

1. *Reporter*, Special Issue, 3 August 1997, 7, 13. At a hearing on the tenth anniversary of SMCRA, Arizona representative Morris Udall, who had played an instrumental role in passing the law, said the act was "fundamentally sound," though hampered by a few states with weak regulatory programs, the change in regulatory philosophy when President Ronald Reagan appointed James Watt as secretary of interior, and a few recalcitrant coal operators. American Mining Congress representative Ben E. Lusk called SMCRA's standards "demanding, inflexible, often counterproductive and always costly" but also described the law as "fundamentally . . . sound." House Subcommittee on Energy and the Environment, Committee on Interior and Insular Affairs, *Tenth Anniversary of the Surface Mining Control and Reclamation Act of 1977*, 1, 65–66.

2. Bruce Daniel Rogers, "Public Policy and Pollution Abatement: TVA and Strip Mining," (Ph.D. diss., Indiana University, 1973), 184–91.

3. Tennessee, *Public Acts*, 1967, p. 88, 95–96.

4. *Oak Ridger*, 7 July 1971, 3; *Nashville Tennessean*, 16 February 1972, 1, 2; 20 February 1972, 1, 2; 22 February 1972, 10.

5. *Nashville Tennessean*, 8 March 1972, 1, 5; 9 March 1972, 1, 4; 15 March 1972, 1; 16 March 1972, 1.

6. Ibid., 12 September 1971, 3; 13 September 1971, 8.

7. Ibid., 12 September 1971, 1, 2; 16 September 1971, 9.

8. *The Mountain Has No Seed*, Broadside/Appalachian Video Network, Archives and Special Collections, East Tennessee State University.

9. *Oak Ridger*, 30 September 1971, 4; *Nashville Tennessean*, 19 September 1971, 24; J. W. Bradley interview. A press release listed the complainants as follows: Mr. Fred Jones, Briceville, machinist; Mr. and Mrs. Doyle Burns, White Oak, superintendent of machine shop and machine operator; Miss Marie Cirillo, Clairfield, community development; Mrs. Vercie Norton, Duff, textile worker; Mr. Millard Ridenour, White Oak, retired miner; Mr. James S. Hatmaker, Eagan, equipment operator; Mr. Clarence Hackler, Clairfield, truck driver; Mr. J. W. Bradley, Petros, electrician; Mr. Ronnie H. Beck, Coalfield, instrument mechanic; Mr. Bill E. Christopher, Petros, instrument mechanic; Mr. Sherman Fetterman, Oneida, college student; Mr. Cedric Jurgens, Oneida, retired Marine officer. Research for the complaint was done in the summer of 1971 for the Vanderbilt Student Health Coalition by John Gaventa, Ellen Ormond, Bob Thompson, Professor Lester Salamon, and Heleny Cook, a Sarah Lawrence College student. Press release, 16 September 1971, copy in author's possession, private papers of Richard Cartwright Austin.

10. J. W. and Kate Bradley returned to their home in May 1977, after a decision by a chancery court judge that included a payment of $2,700 by the company to the Bradleys, for damages to the property and denial of access to the land; J. W. Bradley interview; *Mountain Life and Work* 53 (August 1977): 35; Bill Allen, "Save Our Cumberland Mountains: Growth and Change within a Grassroots Organization," cited in *Fighting Back in Appalachia: Traditions of Resistance and Change*, ed. Stephen L. Fisher (Philadelphia: Temple University Press, 1993), 86–87.

11. In 1975, the State Equalization Board ruled on the original SOCM complaint in favor of the organization. Thereafter, all tracts of coal land, whether they were active mine sites or not, were to be taxed as commercial property, at 40 percent rather than the old assessment of 25 percent for farmland. Three years later, SOCM petitioned county tax equalization boards to raise assessments and were fairly successful. Anderson County raised their total assessment $1.5 million, adding $90,000 in new taxes. The group also succeeded in changing the classification of "economic"—and thus taxable—from 36 or more inches to 24 or more inches of coal. Operators continued to challenge the state board's ruling, however, and it was not until 1979 that it issued a judgment on an appeal of the Coal Creek Mining and Manufacturing Company, upholding its earlier decision that undeveloped mineral reserves must be taxed. *SOCM Sentinel* (February 1975): 1; *SOCM Sentinel* (April 1978): 1; *SOCM Sentinel* (April 1979): 1; *SOCM Sentinel* (March 1975): 5; *Mountain Life and Work* 49 (May 1973): 16; Coleman McCarthy, *Disturbers of the Peace: Profiles in Nonadjustment* (Boston: Houghton Mifflin, 1973), 165.

12. *SOCM Sentinel* (March 1978): 1; *SOCM Sentinel* (April 1978): 5.

13. *SOCM Sentinel* (February 1975): 3; *SOCM Sentinel* (March 1975): 1–2; *SOCM Sentinel* (January 1976): 1.

14. *Mountain Life and Work* 52 (January 1976): 4–6; *SOCM Sentinel* (July 1975): 3; *SOCM Sentinel* (October 1975): 1–2; *SOCM Sentinel* (November 1975): 3; *SOCM Sentinel* (January 1976): 5; Allen, "Save Our Cumberland Mountains," 87.

15. Allen, "Save Our Cumberland Mountains," 90; *SOCM Sentinel* (October 1978): 3; *SOCM Sentinel* (August 1979): 1; *SOCM Sentinel* (September 1979): 2; *SOCM Sentinel* (October 1979): 4.

16. Allen, "Save Our Cumberland Mountains," 92.

17. Timothy D. Berry, "Pursuit of Democracy: An Organizing History of Save Our Cumberland Mountains (SOCM), 1972–1992" (Master's thesis, University of Kentucky, 1996), 125–26.

18. Melanie A. Zuercher, ed., *Making History: The First Ten Years of KFTC* (Prestonsburg: Kentuckians for the Commonwealth, 1991), 1; *Courier Journal*, 8 April 1977, 1.

19. *Mountain Life and Work* 53 (August 1977): 29.

20. Jim Schwab, *Deeper Shades of Green: The Rise of Blue-Collar and Minority Environmentalism in America* (San Francisco: Sierra Club Books, 1994), 290–91.

21. Zuercher, *Making History*, 2–3.

22. Ibid., 9–11; Beth Spence to Alliance Members, 28 June 1977, Folder "Strip Mining," Boxes 1/2, West Virginia and Regional History Collection, West Virginia University (hereafter cited as WVRH Papers); *Mountain Life and Work* 53 (August 1977): 46; *Mountain Life and Work* 54 (January 1978): 23.

23. *Mountain Life and Work* 54 (August 1978): 13–14; *SOCM Sentinel* (March 1978): 8.

24. *SOCM Sentinel* (August 1979): 4.

25. Zuercher, *Making History*, 11–13.

26. Ibid., 13–17.

27. Schwab, *Deeper Shades of Green*, 294.

28. Ibid., 298–99.

29. Ibid., 298–99.

30. *Lexington Herald*, 2 August 1987, 1, 14.

31. *Charleston Gazette*, 3 May 1998; *Charleston Gazette*, 31 March 1998.

32. Ted Williams, "Mountain Madness," *Audubon Magazine* (May/June): 36; Maryanne Vollers, "Razing Appalachia," *Mother Jones* 24 (July/August 1999): 36–43, 86–87. Since 1987, Wyoming has been the leading producer of coal in the nation. Kentucky is second in terms of coal production and West Virginia is third. Office of Surface Mining, Department of the Interior, *20th Anniversary: Surface Mining Control and Reclamation Act, A Report on the Protection and Restoration if the Nation's Land and Water Resources under the Surface Mining Law*, Part II (1999), 72.

33. *Charleston Gazette*, 3 May 1998; *Charleston Gazette*, 9 August 1998; Vollers, "Razing Appalachia."

34. Vollers, "Razing Appalachia"; *Charleston Gazette*, 6 June 1999.

Conclusion

1. *Hazard Herald*, 26 August 1965, Folder 4, "Strip Mining, January 10, 1965–December 23, 1965," Box 19, Harry Caudill Papers, Special Collections, University of Kentucky.

2. *People Speak Out on Strip Mining* (Berea, Ky.: Council of Southern Mountains, 1971), n.p.

3. *Courier Journal*, 6 July 1967, Folder 6, Box 5, Gordon Ebersole/The Congress for Appalachian Development Manuscript Collection, East Tennessee State University.

4. "To Look Over the Land and Take Care of It," Broadside/Appalachian Video Network, Archives and Special Collections, East Tennessee State University.

5. The argument that the opposition played a crucial role in the enactment of control laws, by causing mayhem and pressuring lawmakers for a ban, runs contrary to the conclusion of a study of Kentucky's regulatory legislation. "The transient quality of the protests," Marc Landy argues, "couple with the political insignificance of the protestors, meant that the Governor could have ignored them had he wished to do so. Instead he recognized that they presented an excellent occasion for launching his strip mine initiative, and he therefore decided to identify himself personally with the protestors' demands." Marc Landy, *The Politics of Environmental Reform: Controlling Kentucky Strip Mining* (Washington, D.C.: Resources for the Future, 1976), 8.

6. Mary Beth Bingman, "Stopping the Bulldozers: What Difference Did It Make?" in *Fighting Back in Appalachia: Traditions of Resistance and Change*, ed. Stephen L. Fisher (Philadelphia: Temple University Press, 1993), 29.

selected bibliography

Manuscript Collections

Athens, Ohio
Special Collections, Ohio University
    United Mine Workers of America, District 6, Papers

Berea, Kentucky
Southern Appalachian Archives, Berea College
    Appalachian Volunteers, 1963–1970 Records

Blackey, Kentucky
Joe and Gaynell Begley Papers (private collection)

Columbus, Ohio
Ohio Historical Society
    Thomas J. Herbert Papers
    Frank J. Lausche Papers
    William O'Neill Papers

Dungannon, Virginia
Richard Cartwright Austin Papers (private collection)

Huntington, West Virginia
Special Collections, Marshall University
    Ken Hechler Manuscript Collection

Indiana, Pennsylvania
Special Collections, Indiana University of Pennsylvania
    John P. Saylor Manuscript Collection
    United Mine Workers of America Papers

Johnson City, Tennessee
Archives of Appalachia, East Tennessee State University
    Gordon Ebersole/The Congress for Appalachian Development Manuscript
    Collection

Lexington, Kentucky
Special Collections, University of Kentucky
   Harry Caudill Manuscript Collection

Morgantown, West Virginia
West Virginia University
   West Virginia and Regional History Collection

University Park, Pennsylvania
Historical Collections and Labor Archives, Pennsylvania State University
   William Scranton Papers
   United Mine Workers of America Papers

Interviews and Oral History Collections

Austin, Richard Cartwright. Interview by author. Tape recording and handwritten notes. Dungannon, Virginia, 15 September 1998.

Begley, Joe. Interview by author. Tape recording and handwritten notes. Blackey, Kentucky, 14 September 1998.

Bradley, J. W. Interview by Joshua Low by telephone. Tape recording. 30 October 2001.

Carawan, Guy and Candie, eds. *Voices from the Mountains*. New York: Alfred A. Knopf, 1975.

Dunlap, Louise. Interview by author. Telephone, August 1999.

Hechler, Ken. Interview by author. Telephone, 8 September 1998.

Kilgore, Frank. Interview by author. Tape recording and handwritten notes. St. Paul, Virginia, 14 September 1998.

Kobak, Sue Ella. Interview by author. Tape recording and handwritten notes. Pennington Gap, Virginia, 15 September 1998.

War on Poverty in Appalachian Kentucky Oral History Project. Special Collections, University of Kentucky.

Newspapers and Periodicals

Appalachian Strip Mining Information Service (West Virginia)
Buckeye Farm News (Ohio)
Cadiz Republican (Ohio)
Charleston Gazette (West Virginia)
Citizens to Abolish Strip Mining, Inc. (West Virginia)
Coal Age
Columbus Dispatch (Ohio)
Courier Journal (Louisville, Kentucky)
Daily Enterprise (Harlan, Kentucky)
Fed-O-Gram (Ohio)
Hazard Herald (Kentucky)

Lexington Herald (Kentucky)
Mining Congress Journal
Mountain Eagle (Whitesburg, Kentucky)
Mountain Life and Work
Nashville Tennessean
New York Times
Oak Ridger (Tennessee)
Pittsburgh Press (Pennsylvania)
Plain Dealer (Cleveland, Ohio)
Reporter (Citizens' Coal Coalition)
Sierra Club Bulletin
SOCM Sentinel (Tennessee)
SOK Newsletter (Kentucky)
Strip Mining BULLETIN (Kentucky)
Times Leader (Martins Ferry and Bellaire, Ohio)
United Mine Workers Journal
World-News (Roanoke, Virginia)

Proceedings

Journal of the Proceedings of the Ohio State Grange.
People Speak Out on Strip Mining. Berea, Ky.: Council of Southern Mountains, 1971.
Proceedings of the Symposium on Strip Mine Reclamation, June 23–25, 1965
[Kentucky].
United Mine Workers of America. *Proceedings of the 47th Constitutional Convention,*
1976.
———. *Proceedings of the 46th Constitutional Convention, 1973.*

Government Documents

Boccardy, Joseph A., and Willard M. Spaulding Jr. *Effects of Surface Mining on Fish*
and Wildlife in Appalachia. Washington, D.C.: U.S. Department of Interior, 1968.
Bownocker, J. A., and Ethel S. Dean. *Analyses of the Coals of Ohio.* Geological Survey
of Ohio, 4th Series, Bulletin 34, 1929.
Collier, Charles. "Influences of Strip Mining on the Hydrologic Environment of
Parts of Beaver Creek Basin, Kentucky, 1955–59." U.S. Geological Survey Profes-
sional Paper 427-B. Washington, D.C.: Government Printing Office, 1964.
Congressional Record. 89th Cong., 2d sess., 1966. Vol. 112
Congressional Record. 87th Cong., 2d sess., 1962. Vol. 108.
Council on Environmental Quality. *Coal Surface Mining and Reclamation: An Envi-*
ronmental and Economic Assessment of Alternatives. Washington, D.C.: Govern-
ment Printing Office, 1973.
Dyer, Kenneth L., and Willie R. Curtis. *Effect of Strip Mining on Water Quality in*
Small Streams in Eastern Kentucky, 1967–1975. USDA Forest Service Research
Paper NE-372, 1977.

Indiana. *Acts* (1941).

Kentucky Strip Mining and Reclamation Commission. *Strip Mining in Kentucky.* Frankfort, Ky.: 1965.

Kentucky. *Acts* (1966, 1964, 1954).

Mathematica, Inc. *Design of Surface Mining Systems in the Eastern Kentucky Coal Fields.* Vols. 1 and 2. Frankfort, Kentucky: Department of Natural Resources and Environmental Protection, 1974.

Moore, H. R., and R. C. Headington. *Agricultural and Land Use As Affected by Strip Mining of Coal in Eastern Ohio.* Ohio State University Department of Rural Economics, Bulletin No. 135, n.d.

Ohio. *Acts* (1965).

Ohio. *Laws* (1947).

Pennsylvania. *Laws* (1945, 1961).

President's Appalachian Regional Commission. *Appalachia.* Washington, D.C.: Government Printing Office, 1964.

Report of the Strip Mining Study Commission to the Governor and the 97th General Assembly of the State of Ohio. 1947.

Schmidt, R. A., and W. C. Stoneman. *A Study of Coal Surface Mining in West Virginia.* Menlo, Calif.: Stanford Research Institute, 1972.

Tennessee Department of Conservation and Commerce and Tennessee Valley Authority. *Conditions Resulting from Strip Mining for Coal in Tennessee.* 1960.

Tennessee. *Public Acts* (1967).

U.S. Bureau of Mines. *Statistical Appendix to Minerals Yearbook, 1932–1933.* Washington, D.C.: Government Printing Office, 1934.

———. *Minerals Yearbook: 1939.* Washington, D.C.: Government Printing Office, 1939.

———. *Minerals Yearbook: Review of 1940.* Washington, D.C.: Government Printing Office, 1941.

———. *Minerals Yearbook: 1948.* Washington, D.C.: Government Printing Office, 1950.

———. *Minerals Yearbook: 1955, Fuels.* Washington, D.C.: Government Printing Office, 1958.

———. *Minerals Yearbook: Fuels, 1963.* Washington, D.C.: Government Printing Office, 1964.

———. *Minerals Yearbook, 1973.* Vol. 1, *Metals, Minerals, and Fuels.* Washington, D.C.: Government Printing Office, 1975.

———. *Minerals Yearbook, 1976.* Vol. 1, *Metals, Minerals, and Fuels.* Washington, D.C.: Government Printing Office, 1978.

U.S. Department of the Interior. *Study of Strip and Surface Mining in Appalachia: An Interim Report to the Appalachian Regional Commission.* Washington, D.C.: Government Printing Office, 1966.

U.S. Department of the Interior, Office of Surface Mining. *20th Anniversary: Surface Mining Control and Reclamation Act, A Report on the Protection and Restoration if the Nation's Land and Water Resources under the Surface Mining Law.* Part II. Washington, D.C.: Government Printing Office, 1999.

U.S. House Committee on Interior and Insular Affairs. *Report to Accompany H.R. 6482.* 92d Cong., 2d sess., 1972.

———. *Surface Mining Control and Reclamation Act of 1974, Report . . . to Accompany H.R. 11500.* 93rd Cong., 2d sess., 1974.

———. *Surface Mining Control and Reclamation Act of 1976: Report . . . to Accompany H.R. 9725.* 94th Cong., 2d sess., 1976.

———. *Surface Mining Control and Reclamation Act of 1976: Report . . . to Accompany H.R. 13950.* 94th Cong., 2d sess., 1976.

———. *Surface Mining Control and Reclamation Act of 1976: Report . . . to Accompany H.R. 9725.* 94th Cong., 2d sess., 1976.

U.S. House Subcommittee on Energy and the Environment. *Hearings before the Subcommittee on Energy and the Environment of the Committee on Interior and Insular Affairs on H.R. 2, Surface Mining Control and Reclamation Act of 1977.* 95th Cong., 1st sess., 1977.

U.S. House Subcommittee on Energy and the Environment and Subcommittee on Mines and Mining. *Hearings before the Subcommittee on Energy and the Environment and the Subcommittee on Mines and Mining of the Committee on Interior and Insular Affairs on the President's Veto of H.R. 25.* 94th Cong., 1st sess., 1975.

U.S. House Subcommittee on Energy and the Environment, Committee on Interior and Insular Affairs. *Tenth Anniversary of the Surface Mining Control and Reclamation Act of 1977.* 100th Cong., 1st sess., 1987.

U.S. House Subcommittee on the Environment and Subcommittee on Mines and Mining. *Hearings before the Subcommittee on the Environment and Subcommittee on Mines and Mining of the Committee on Interior and Insular Affairs on H.R. 3 and Related Bills, Relating to the Regulation of Surface Mining.* 93rd Cong., 1st sess., 1973.

U.S. House Subcommittee on Mines and Mining. *Hearings before the Subcommittee on Mines and Mining of the Committee on Interior and Insular Affairs . . . on H.R. 60 and Related Bills, Relating to Strip Mining.* 92d Cong., 1st sess., 1971.

U.S. House and Senate Committee of Conference. *Surface Mining Control and Reclamation Act of 1975: Report . . . to Accompany H.R. 25.* 94th Cong., 1st sess., 1975.

U.S. Senate Committee on Interior and Insular Affairs. *Committee on Interior and Insular Affairs, Surface Mining Control and Reclamation Act of 1975: Report . . . to Accompany S. 7.* 94th Cong., 1st sess., 1975.

———. *Committee on Interior and Insular Affairs, Surface Mining Control and Reclamation Act of 1977: Conference Report . . . to Accompany H.R. 2.* 95th Cong., 1st sess., 1977.

———. *Hearings before the Committee on Interior and Insular Affairs on S. 425 . . . [and] S. 923.* 93rd Cong., 1st sess., 1973.

———. *Hearings before the Committee on Interior and Insular Affairs on S. 3132, S. 3126, and S. 217, Bills to Provide for Federal-State Cooperation in the Reclamation of Strip Mined Lands.* 90th Cong., 2d sess., 1968.

———. *Surface Mining Reclamation Act of 1972: Report . . . to Accompany S. 630.* 92d Cong., 2d sess., 1972.

————. *Surface Mining Reclamation Act of 1973, Report to Accompany S. 425.* 93rd Cong., 1st sess., 1973.

U.S. Senate Subcommittee on Minerals, Materials, and Fuels. *Hearings before the Subcommittee on Minerals, Materials, and Fuels of the Committee of Interior and Insular Affairs Pursuant to S. Res. 45, A National Fuels and Energy Policy Study, on S. 77 . . . S. 2777.* 92d Cong., 1st sess., 1971–1972.

U.S. Senate Subcommittee on Public Lands and Resources. *Hearing before the Subcommittee on Public Lands and Resources of the Committee on Energy and Natural Resources on the Implementation of Public Law 95-87, the Surface Mining Control and Reclamation Act and Pending Legislation to Increase Authorization of Appropriations.* 95th Cong., 2d sess., 1978.

index